Progress

*Published in cooperation with
the Center for American Places,
Santa Fe, New Mexico, and
Harrisonburg, Virginia*

PRGRESS

*Geographical
Essays*

Edited by Robert David Sack

The Johns Hopkins University Press | **Baltimore and London**

© 2002 The Johns Hopkins University Press
All rights reserved. Published 2002
Printed in the United States of America on acid-free paper

9 8 7 6 5 4 3 2 1

The Johns Hopkins University Press
2715 North Charles Street
Baltimore, Maryland 21218-4363
www.press.jhu.edu

Library of Congress Cataloging-in-Publication Data

Progress : geographical essays / edited by Robert David Sack.
    p. cm.
  Includes bibliographical references and index
  ISBN 0-8018-6871-8 (hardcover : alk. paper) — ISBN
0-8018-6872-6 (pbk. : alk. paper)
    1. Progress—Congresses. 2. Geography—Congresses.
3. Human geography—Congresses. I. Sack, Robert David.
HM891 .P76 2002
303.44—dc21

                                    2001005479

A catalog record for this book is available from the British
Library.

Contents

Introduction

The essays in this volume were originally presented as part of a symposium on the topic of progress held at the University of Wisconsin, Madison, in May 1998, on the occasion of Professor Yi-Fu Tuan's retirement from active teaching. The university wanted to honor Professor Tuan for his outstanding contributions to geography and to our understanding of the human condition. We chose the topic of progress because it is complex and contested, because it is important to Western culture and science, and because it can benefit from a geographical perspective.

The connection between geography and progress is fundamental. This is clear in the age-old definition of geography as the understanding of how human beings transform the world and make it habitable. Transforming is about changing, and this immediately raises issues of progress. The connection deepens when we consider two qualities of this definition. One is its suggestion that all human beings are geographical agents, for we are all engaged in this transformation. Geography is not only something we study but also something we create. We are geographical beings. The second quality is the suggestion that changes resulting from human processes are different from changes brought about by other agents. That is, we differ from other animals in the way we affect and alter the world. If this were not the case, then why would our activities merit special attention? The differences, as well as their relation to progress, can be better understood by unpacking our definition of geography—as the transformation of the world to make it habitable—into the following three-part sequence. First, we transform the world because humans are constitutionally incapable of accepting reality as it is. This means that we are always transforming it. This is equally true for the reality of both the natural and the cultural world. We cannot simply live out in the open, exposed to nature as it really is; we need to construct shelters and organize the world into areas for living, learning, and working. Once we create this new cultural reality, we alter that too. So we transform reality, then transform the reality we have transformed into a new reality, and so on. Second, we do this transforming by constructing places. Places are the geographical instruments that allow us to transform nature and culture, to combine and interweave the two, and, more generally, place allows us to undertake projects. As geographical beings we are place-makers. Third, we transform reality and make places because we have an image of what we think reality can and ought to be. We imagine realities that have not yet existed and try to create them.[1]

This unpacked definition of geography not only sets us apart from other animals but leads to a wide range of interpretations of progress. Whether our project is clearing a forest to create a settlement, damming a stream to produce electricity, or replacing cleared farmland with a shopping mall, we can think of progress as the stages in that project. How much progress have we made in clearing the land, in building the dam, in erecting the shopping mall? Progress in this sense can be defined very narrowly as the distance we need to go, or the stages we must pass through, to complete a project.

But we can also have progress mean that these are worthy projects, projects that ought, or ought not, to take place. Either way, we are now appealing to a more general sense of the direction human transformations should take. These more general issues seem to invoke the idea that progress is a moral issue. There are morally better and worse transformations and types of places. Some would say that the moral thing to do is to avoid clearing the land, damming the stream, and erecting the shopping mall. According to this view, we should keep things as they are or even return to some former and perhaps simpler, more "natural" state. The moral becomes even more pronounced when the claim is made that we must change some things in order to create an environment in which morality can flourish. As these and other examples discussed in this collection of essays suggest, a very wide range of views about progress is invoked by, and intimately linked to, our geographical nature as earth transformers and place-makers.

This range has important historical dimensions. Our definition of geography is age-old, but before the modern period—before, let us say, the rise of capitalism and the age of exploration, the period around A.D. 1500—the extent and scope of our transformations at any particular place and time were far more gradual and narrower, and there did not exist a picture of how these effects accumulated over human history. That is, before modernity people were not very aware of the power of human geographical agency, and indeed, it was far less powerful than it is now. This limited effect and limited awareness lent to the premodern a different range of meanings for progress than exists now, and these differences help distinguish the premodern from the modern.

Even the most complex premodern cultures were bound by tradition, which was often embedded within a pervasive religious view of the world. The world, including the landscape and one's own culture, with its political organization and stations of life, was generally believed to have been divinely created. Most everything was under divine authority and served a divine purpose. Things were understood in functional and teleological terms. Activities provided functions that served the general purpose of design. By and large, life in such societies did not change rapidly. People were born into particular occupations and classes. They could make progress by acquiring skills and learning the technique of crafts or projects demanded of their station, but their place in society and the social structure in general did not become the focus of change. After all, God (or the gods) had control over events, and to drastically change them would seem a sacrilege. Those changes that did occur were most often justified by appeals to tradition or custom. A change might be needed because a project was believed to have deviated from its traditional form. The change would then return it to the way it originally was, which was the way it should be. Since the customs and practices were often couched in religious terms, a change back to the original returned society to a purer form. Thus, change was a form of purification. This idea, along with a more sluggish system of technological change and an agricultural society driven by rhythms and cycles of nature, presented time itself as a cycle; we may have strayed from the starting point (as in the creation of the earth or the creation of the kingdom), but we tried to return to it over and over again. Radical revolutions were out of place in such societies. Peasants might rise up, but their revolutions were not intended to abolish the foundations of these socioeconomic systems. Rather, those in revolt wished either not to be peasants or to return to the way peasant society ought to be.

An important exception to this cyclical sense of time was the Jewish, and especially early Christian, idea that soon a cataclysmic event would occur that would lead to redemption. Still, even in the early years of Christianity believers saw little that could be done socially or politically to hasten the end.

They believed that individuals should simply prepare their own personal lives for redemption. By the early Middle Ages the "end" had been pushed far into the future, and the indefinite time remaining had been placed within a rigidly prescribed order and hierarchy bound together by custom and by cycles of good and bad times. People could enter a new world—a paradise—after they died, but worldly events would remain virtually the same, perhaps repeating minor cycles of good and bad times, as had been the case in most other premodern societies. So even in Christian societies progress at the social level did not loom large, and progress at the individual level existed within the narrow range of developing talents and skills for undertaking projects that had mostly been undertaken before.

This situation changed with the rise of modernity. Now (and I will speak of this in the present tense, for we are part of this period) it seems that time brings with it a continuously changing world. We expect the world to be different in the future than it is now, and it certainly is different now than it was. We have become not only more powerful geographical agents but also more aware of our effects. So time now brings changes of all sorts. Accompanying our increased power has been the development of science and technology, which helped accelerate change and also pushed aside the idea of a religiously designed and purposeful world broadly based on functional and teleological explanations. As these have been pushed into the background (though not, as we shall see, entirely) and as our powers have increased, it has become increasingly difficult to evaluate what these changes mean. Are we making progress or not? And what kind of progress? Some may argue that moving away from traditional society, with its narrow and rigid world-view and stultifying social environment, is itself a move in the right direction. Some others seem to have transferred their blind faith from religion to change itself: the future will somehow be not only different but better than the past, and this betterment will continue. Others have a more complex view, thinking of only some changes as progress and developing different and more nuanced meanings of the term. Still others have come to see virtually any change at all as a danger, arguing that progress would involve returning to what (they think) once was.

Two general strands run through these interpretations. First, among the many contemporary meanings of *progress* is still the old, simple, and probably universally shared sense associated with acquiring skills and undertaking tasks to complete a project. This idea of making progress does not say very much else, but it does provide everyone with a basic sense of progress. Even children experience progress in this sense as they learn to crawl and then to stand upright and as they learn to use language. Second, in the modern period, no matter how we define progress, we see ourselves as the primary source. We are more powerful geographical agents and we are more aware of our role. It is therefore all the more pressing that we understand what progress does and should mean.

This collection of essays is about the relationship between geography and progress. By no means can six essays cover all there is to say on this enormous topic, but they do provide a range. Each essay stands on its own, so they do not have to be read in sequence, but their arrangement does offer a progression, as we shall see. An important point about modernity is that contemporary science rarely thinks of nature in terms of design. Natural-science theories in general and Darwinian evolutionary theory in particular see nature as forces that cause things to happen but lack an overall direction or purpose. Nature is constantly changing, but, generally speaking, it has no preferred direction or goal. Progress as an improvement or a move toward a goal is difficult to find. The first essay in the collection, Thomas Vale's "From Clements and Davis to Gould and Botkin: Ideals of Progress in Physical Geography," takes up this issue of progress or its lack in the natural realm. Vale examines models in four areas of physical geography (vegetation, soils, landforms, and climate) to determine whether they see nature as progressing in either or both of two ways—following linear, directional pathways and moving toward a higher and better form. Here Vale is employing Stephen Gould's taxonomy of ideas of progress in science. Following Gould, Vale adds that progress in either or both senses also assumes that changes in the physical world "adapt" in a precise way to the environment, that they reflect a tight causality or determination, and that they are gradual. Gould argues that overall the natural world exhibits directionless change without betterment (so that there is no progress) and that the changes are haphazard (as opposed to finely adaptive), not strictly determined, and not gradual but rather rapid and pulsed. Which set of ideas is found in the geographical studies of vegetation, soils, landforms, and climate? Vale finds the strongest commitment to progress, though never one that goes unchallenged, in models of vegetation, weaker commitments in soil models and in some landform ones, and scarcely a vestige of commitment in

---

climate models. Vale offers several explanations for these differences, among them the general time frame and pace of change that each examines and the closeness of the models to those of physics. Seeing progress in natural systems has serious implications for how we as humans understand our role in transforming or preserving nature. If nature prefers some states to others, should we too?

The second essay focuses on social systems and describes a fundamental change in a set of meanings of progress that occurred with the rise of modernity. Ken Olwig, in "Landscape, Place, and the State of Progress," uses changes in political and social representations of state power from Tudor to Stuart England to illustrate more general transformations in the idea of progress. It was the tradition in England at Elizabeth's time and before for a monarch to visit the various provinces and counties that constituted the realms of the kingdom. These visits, or rounds, constituted a royal progression and were a means by which the various communities, as parts, recognized the power of the monarch, and the monarch recognized the local and distinctive parts of the realm, with rights and customary laws. This political progression was circuitous and cyclical, as was the general idea of progress at the time, and conformed to a world of distinct places with distinct customs and traditions. Not only was progress circuitous but the term *landscape* referred to these places of local authority and power. This was to change.

In the court of James I of England there occurred a shift in viewpoint from a kingdom of local and distinct places or landscapes, each with its own customary laws, to a kingdom of space in which the particular places or landscapes were now conceived as a backdrop to the power and law that emanated from the king and court. Places became conceptually "thinned out," so that *landscape* now meant scene or background; local and customary law was superseded by a more universal sense of law, and progress was no longer circuitous but became a linear and continuous transformation of things in space. Olwig's essay shows how this new conception of landscape, space, and progress was introduced in and elaborated through courtly theater—theater that employed the new techniques of perspective painting to portray the country as a space and background against which the monarch viewed the enactment of his power on the land. Progress is now linear and directional, and landscape becomes scenes and stages in this progress. Progress here means change for the better, and, as Olwig shows, it persists into our world, where many planners still see the thinning out and destruction of distinctive places and the emphasis on space a mark of progress.

The modern period thus ushers in a sense of change that is linear and seen by most as positive or progress. Moreover, this change is associated with changes in the types of place we construct; places are now thinned out and encourage a system of flows through space. So the modern world becomes a world of ever greater transformations, a world in which movement through space vies in importance with being situated in place. The next two essays are concerned with how, most recently, these geographical transformations have caused disenchantment and anxiety. "The Disenchanted Future," by David Lowenthal, considers the causes and consequences of contemporary erosion in confidence about progress. Awareness that humans themselves can change and that they can also change the world leads on the one side to an unbridled optimism in the possibility of human perfectibility here on earth, or, if not perfectibility then at least human betterment. On the other side, it leads to anxieties resulting from doubts about whether we can ever know and agree on what is good and to anxieties that stem from the fact that despite these uncertainties, we are continuing to change the world socially, intellectually, and environmentally. We now fear that our own activities will lead to a global and cataclysmic end.

This fear leads many to desire an escape to a nostalgic past that not only is socially simpler and static but contains a nature that is "purer" and to which we can be closer. Lowenthal sees several more specific forebodings about ourselves that lead to a gloomy conception of the future. Most now have a fear of what is unknowable in part because they are convinced that the unintended consequences we set in motion will almost always be destructive. We have become involved in decision-making processes that focus on the immediate and momentary rather than on the long-term, and the media encourage this. We no longer trust experts, and we also mistrust the power of education to enlighten.

As a corrective, Lowenthal would like to welcome change, which is inevitable anyway, and discourage the notion that the clock can be turned back. Change should not, however, mean a severing with the past but rather an incorporation of it. We should stress stewardship, which would make our concerns less focused on the present and the immediate, and we should never stop anticipating delight, even in future possibilities.

Yi-Fu Tuan, in his essay "Progress and Anxiety," also takes up the problem of our unease with progress. Every human group has a sense of progress, at least insofar as a project or indeed the entire culture could only happen as a cumulative process. Yet this very idea creates anxieties because in pointing to change it reminds us of uncertainty and conditionality. Permanence, and with it a lack of significant choice, can make us feel secure. Change makes us feel less certain and opens up a realm of choice. These tensions are found in the anxiety of growing up, where the child, in taking his or her first step, opens up a new world of choice while teetering on unsteady feet. They are found in anxiety over material progress, where even the simplest society may succeed in chopping down trees to clear a field for crops, only to become anxious about whether the crops will grow, and the forest return. Even if we succeed, we may worry that we are rising above our own station in the cosmic scheme of things, which will offend the gods or cause bad luck. And if we do not worry about this, we will be anxious because as we come to know and control more, we also realize how very much more is still unknown, and uncontrolled. Even moral progress creates anxieties. We may believe that it is morally better to see ourselves as part of humanity rather than as simply part of a family or tribe. Expanding our sense of obligations and care for distant others is a move in the right direction. Yet such moral progress is also a source of anxiety. Before, when our concerns were about family or tribe, we knew what to do, whereas now that we may feel concern for many, or all others, we are overwhelmed by opportunities and choice. Whose needs are more important, and what is best for each? As our connections and responsibilities expand in a global system, we become overwhelmed by choices and obligations that are themselves changing and expanding.

The last two essays focus on moral progress. In "Perfectibility and Democratic Place-Making" J. Nicholas Entrikin discusses the idea of moral improvement or progress, often called moral *perfectibility*, and its connection to democratic places. Perfectibility is often associated with order and symmetry in life and in one's surroundings. Stability and order are then found in landscapes inspired by perfectibility. In contrast, democratically organized places are often subject to shifts in opinions and coalitions. There are tensions between the two, and they may even seem incompatible when taken to the extreme. When perfectibility is not correctly understood—when we have a wrong and intolerant vision of what is good—the result may be a tyrannical control over place. On the other hand, a radically democratic system of places, in which multiple and conflicting voices must be not only heard but accommodated, might produce an incoherent landscape with few projects undertaken and little possibility for improvement. Entrikin sees a blending of the two if we consider each more realistically. Perfectionism is never attainable, but it can be a healthy ideal if it means the continuous act of self-discovery and self-improvement. This type of perfectionism must be embedded, moreover, in a democratic way of life. But the democratic here is not simply a process that assures that everyone's self-interest is respected and preserved but rather a system that helps us become better individuals by being open to new and better ideas—by living a life of learning, a life of liberal education. In this way democracy's structure of free and open access to and exchange of ideas becomes less a device for assuring our own interests and more a means by which we ourselves can be open to change and improvement. This conjunction of democracy (and its educational function) and perfectibility requires a landscape of democratic and public places.

In the last essay, "Geographical Progress toward the Real and the Good," I take up several of these themes of perfectibility and moral progress. Since our geographical activity continually changes reality, we need to judge whether these changes are for the better. Do they create good or bad places? I argue that there are two ways of doing this that stem from two kinds of judgments about place. The first, called *instrumental geographical judgments*, evaluates the merits of a place in terms of its effectiveness in furthering a project. If the place is effective, it is contributing instrumentally to the attainment of this goal. Instrumental judgments and instrumental progress occur anytime we learn a skill or undertake a project. This extremely common way of judging has the defect of being relative, for it does not answer the question whether the project itself is good.

The second way of judging, *intrinsic geographical judgments*, leads to moral or intrinsic geographical progress. It avoids relativity because it is based on two facets of the good that are real and independent of any particular project and judges whether place contributes to them. The two facets may be phrased as follows: it is good to be more rather than less aware of reality, and it is good to have a reality that is more rather than less varied and complex. These become regulative principles that, when used jointly, can guide our place-making. As geographical agents we should then create and maintain places that help us become more aware of reality and that help us to create a more varied and complex reality. Places that do so are better than places that do not. We can of course have places that do only one and not both, such as places that increase the variety of the world but diminish our vision, but this is not good. A crack-cocaine house may add to the variety on the landscape while diminishing the awareness of those within. The same is true of a country that builds a wall of censorship around its citizens. It can add to the variety of the world, but it diminishes our awareness—preventing those within from seeing out, and those outside from seeing in. Real or intrinsic geographical progress occurs when we follow intrinsic geographical judgments. Can we make instrumental progress more like intrinsic progress? I believe that we can. Instrumental progress can expose us to qualities of excellence and experiences of beauty that may help us become more aware of the real and the good.

Each of these essays stands alone and does not refer to the others. But they all address in different ways very similar themes about the connections between geography and progress, themes that ultimately have to do with changing the world for the better.

Note

1. The unpacked definition draws on Yi-Fu Tuan's understanding of human beings as constitutionally incapable of accepting reality as it is and therefore continuously escaping to new and imagined worlds (see Yi-Fu Tuan, *Escapism* [Baltimore: Johns Hopkins UP, 1998]).

Progress

Chapter I

From Clements and Davis to Gould and Botkin: Ideals of Progress in Physical Geography

Thomas R. Vale

Somewhere between the extremes of postmodern nihilism and naive positivism lie scientific truths about the natural world. These moderate positions on the middle ground permit truths to reflect the character of the questions asked and the frameworks in which those questions are placed—both of which resonate with the workings of both individual minds and collective cultural and societal contexts—yet at the same time to maintain convictions not only that nature is real but also that our knowledge of it can be genuine. We know, for example, what makes it rain (unequivocally, Western science understands the processes involved in this external reality), although both the scale at which we evaluate condensed water falling from clouds—is the cause of the falling water the differential vapor pressure on minute ice crystals and supercooled liquid droplets, or the mesoscale feature that induces uplift of air, or the variable pattern of insolation that drives atmospheric circulation?—and the assessment of how much of such measured water constitutes "wet" or "dry" may be inextricably linked to human thought.

Progress—as a concept it is ubiquitous in Western thinking. Even in interpretations of the natural world the identification of progressive trends seems commonplace, and that ubiquity raises the question whether such patterns exist more in the natural reality or in the human mind. One contemporary observer, Stephen Jay Gould, gives a clear answer: the ideal of progress has become one of the prejudices of contemporary thinking about evolutionary history. Gould's critique focuses on the human tendency to see erroneously two expressions of progress in evolution—linear, directional pathways and development toward "higher" or "better" forms. Neither expression of progress fits the patterns of actual evolutionary history, according to Gould, who sees instead directionless diversification, which is neither linear nor improving.[1]

Gould adds three other common interpretations of evolution and extends his critique beyond the bounds of evolutionary science to complete what he calls the "four biases of modern science." In addition to (1) the concept of *progress,* Gould identifies (2) *adaptationism,* the sense that everything fits and works in a fine-tuned mechanism; (3) *determinism,* the sense that everything operates for particular results, that is, that tight ties exist between means and ends, processes and results; and (4) *gradualism,* the conviction that temporal change is slow, steady, and constant. By contrast, Gould champions the four alternative characteristics (as I articulate them): (1) directionless change without necessary betterment; (2) haphazard, sloppily tuned mechanisms; (3) multiple ties between process and result; and (4) rapid or pulsed change. In evolution these attributes describe a system of unpredictable biological branching that results from natural selection operating on whatever forms and behaviors might be present at a particular place and time (and thus producing varying results), with the actual diversification occurring sporadically but rapidly in short periods of time that separate long periods during which little change occurs.

This essay explores the extension of Gould's "four biases" into natural history more generally and into physical geography more specifically: How do progress, adaptation, determinism, and gradualism—or their antithetical traits of change that is directionless, haphazard, variable, and pulsed—structure thinking in various parts of physical geography? The discussion begins with vegetation, with each of the two sets of four characteristics described as creating an aggregated world-view. Then it moves on to soils, geomorphology, and climatology. In each exploration I attempt to characterize the major thinking of the subdiscipline in terms of Gould's four biases, their antithetical traits, and the generalized world-views. I present examples of scientific work to illustrate the major thinking; I make no attempt to survey the vast literature in all of the substantive subject areas. This investigation, which finds the several parts of physical geography rather distinctive, concludes with speculations about why such subdisciplinary differences exist.[2]

Vegetation

The scientific study of vegetation dynamics—the patterns of vegetation response to environmental disturbance—began in the early twentieth century with the work of the ecologist Frederick Clements.[3] The Clementsian portrayal of those dynamics conforms to the traits that Gould calls biases (fig. 1.1). First, vegetation is typically envisioned as reflecting environmental conditions, a linkage that suggests an *adaptation* of the plant cover to the environment. Sometimes the adjustment seems precise and fine-tuned; for example, the classic study by Hack and Goodlett in mountainous Virginia reveals an extremely close relationship between vegetation type and topographic position, so much so that the vegetation map mimics accurately the topography.[4] Second, following disturbance of forest vegetation—by natural agents such as fire or human activities such as logging—the subsequent development of the plant cover is described as adhering to the linear pathway, the *progressive* pathway, of secondary succession. Third, succession in forest vegetation remains a vision of *gradual* change—spread over many decades to more than a century—often portrayed as broken into stages from grass, forb, or shrub types, through pole or sapling forest, young forest, mature forest, and finally old growth. Fourth, the end point of the forest's successional sequence, the old growth or climax condition, reflects and is *determined* by the environmental setting; it does not matter where the sequence begins, moreover, because through time the developmental pathway leads to the climax. Taken together, then, the Clementsian model, which reflects each of Gould's four biases, might be described as the *adaptation/progress* perspective.

The second expression of progress—development toward a better state—also typifies successional sequences. The reverence in which climax or old growth forests are held—and the indifference or even disdain expressed for "lesser" stages—suggests that forests improve in qualitative terms as they develop. As we explore other parts of physical geography, we will find that it is only in vegetation dynamics that the linear pathway—one measure of progress—coincides with enhanced virtue—the second measure of progress.

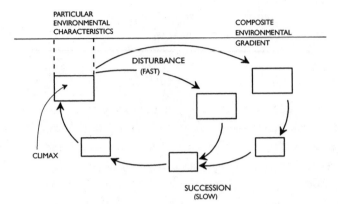

PARTICULAR
ENVIRONMENTAL
CHARACTERISTICS

COMPOSITE
ENVIRONMENTAL
GRADIENT

DISTURBANCE
(FAST)

CLIMAX

SUCCESSION
(SLOW)

Fig. 1.1. Clementsian vegetation dynamics. The horizontal line represents a composite of environmental gradients; boxes represent vegetation conditions or states; arrows between boxes represent change in vegetation over time.

The Clementsian view continues to serve as a general model of vegetation dynamics, although it is often implied or assumed rather than explicitly articulated.[5] Nonetheless, research over many decades has qualified or selectively rejected the details of Clementsian dynamics in particular situations, with the critiques frequently promoting the traits antithetical to the adaptation/progress model (fig. 1.2). First, multiple stable states refute narrow adaptationism; McCune and Allen, for example, found in environmentally identical canyons on the east side of Montana's Bitterroot Mountains forests with variable mixes of tree species, all of which seemed stable and persistent on their individual sites.[6] Second, the sequences that are perceived as succession—each group of species reproducing in the environments created by the preceding group—may reflect processes other than replacement. A prime example, provided by Egler, envisions differential survival—species establishing themselves in a forest stand at the same time but then dying at different times—as creating an impression, however false, of successional reproduction.[7] Third, Egler's *initial floristics* view also contradicts gradualism, because the plants that first occupy a site after disturbance continue to dominate, even if some die out, until the next major disturbance event. Henry and Swan found such a pattern in the three-hundred-year history of a New Hampshire forest, which also revealed a different mix of species—multiple stable states—after each disturbance (fig. 1.3).[8] Fourth, the importance of initial conditions also challenges narrow determinism; for example, Christensen found that on North Carolina's coastal plain whether shortleaf *(Pinus*

Thomas R. Vale

echinata) or loblolly *(Pinus taeda)* pine dominates a site depends upon the timing of farm abandonment and thus of forest establishment—different beginnings yield different ends.[9] As a group, then, the four traits contrary to Gould's biases—multiple states, absence of simple linear development, pulsed change, and influences of initial conditions—stress the critical importance of historical circumstance, of contingency, to vegetation development

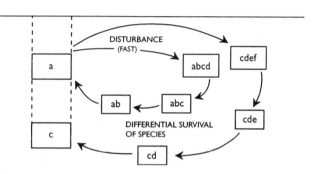

Fig. 1.2. The historical-event interpretation of vegetation dynamics. Small letters in boxes represent biological species; other symbols are explained in the caption to figure 1.1.

Fig. 1.3. Patterns of vegetation change through time identified by J. D. Henry and J. M. A. Swan in a New Hampshire forest.

1665	1665	1665 - 1938
W. PINE	FIRE	W. PINE 1665 - 1687
E. HEMLOCK		E. HEMLOCK 1684 - 1780
SPRUCE		P. BIRCH 1747, 1770, 1795
		BEECH after 1815
		R. OAK 1790 - 1860 die
		FOREST 1907
		W. PINE
		E. HEMLOCK
		BEECH
		P. BIRCH

1938	1938 -
WINDSTORM	E. HEMLOCK
	BLACK BIRCH
	R. MAPLE
	BEECH

after disturbance. Collectively these characteristics might be called—in contrast to *adaptation/progress*—the *historical-event perspective.*

The science of vegetation dynamics has been liberated from a narrow, strict adherence to the adaptation/progress view by an increasing appreciation of the enriching complexity of historical events. This enrichment has taken the form not of a Kuhnian paradigm shift—too much truth resides in the characteristics of succession—but, rather, of a broadening theoretical foundation that encompasses both perspectives.[10]

Biogeographical themes other than vegetation dynamics involve different relationships to the four opposing traits and two aggregated models. Paleovegetation reconstruction, for example, fundamentally depends upon the adaptation/progress perspective. The use of pollen or macrofossil data to recreate environments of the past assumes the existence of a tight link between vegetation and environmental conditions, a link uncomplicated by singular historical events.[11] Dynamics and environmental relationships of animal populations, by contrast, much more strongly than vegetation studies utilize interpretations involving the historical-event perspective.[12]

Soils

Physical geographers interested in soils most likely focus upon soil development as a consequence of environment setting; from such a perspective, the state-factor approach of Hans Jenny becomes central. Jenny formulated the link between characteristics of the soil profile and environmental factors as the *state-factor equation:* soil is a function of climate, relief, parent material, organisms, and time (with additional factors possible in particular situations).[13] In terms of the ideas discussed in this essay, Jenny's expression articulates the adaptation/progress perspective. First, the collective impact of the five factors produces a known and predictable soil profile, an *adaptation* of soil characteristics to environmental setting that directly parallels the similar adaptation of vegetation to the suite of environmental conditions. Second, a recasting of the equation to isolate time as the variable—to say that soil characteristics under conditions of particular climate, relief, parent material, and organisms reflect the age of the soil—yields the model of the chronosequence that is familiar in soils studies. Under these conditions the soil profile develops, or *progresses,* through time, with a simple, linear trend often iden-

Thomas R. Vale

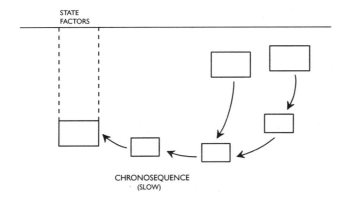

STATE
FACTORS

CHRONOSEQUENCE
(SLOW)

Fig. 1.4. The adaptation/ progress model of soil development. The symbols are explained in the caption to figure 1.1.

tified (fig. 1.4). Third, many soil characteristics, such as movement of clay from an upper to a lower horizon, change at slow rates, an expression of *gradualism*. Fourth, the mix of environmental factors prompts the soil to develop along a singular pathway toward a singular end point; the generalized linkages between soil orders and environmental settings suggest this *determinism* (fig. 1.5).

The definition of progress as a trend toward a "better" state—so obvious in the perceived value of old-growth forest—does not seem apparent in the study of soil's developing along a linear path: no one ever suggests that a well-developed Spodosol or Oxisol embodies greater virtue than an Inceptisol or an Entisol. On the other hand, the soils literature attends disproportionately to well-developed soils. Perhaps the soil scientist interested in genesis sees more to study in such soils and in that way expresses a preference, a qualitative value, for the later states along the soil's developmental pathway.

The adaptation/progress model dominates the study of soils, but the contrary, antithetical views also enliven the literature (fig. 1.6). First, multiple states of soils—under comparable climate, relief, parent material, and organisms—might be generated under different histories (i.e., different expressions of the factor of time). Johnson and Watson-Stegner, for example, described two different soil profiles that might develop in environments that were identical except in their past human land uses.[14] Second, those histories involve not only progressive development of soil characteristics, such as leaching of cations, but also regressive processes, such as tree uprooting

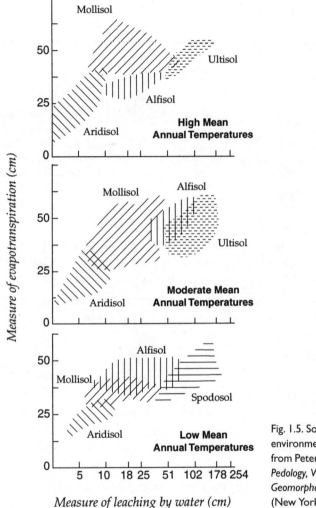

Fig. 1.5. Soil orders in environmental space. Adapted from Peter W. Birkeland, *Pedology, Weathering, and Geomorphological Research* (New York: Oxford UP, 1974).

Measure of leaching by water (cm)

that turns over the surface layers of the soil, activities that counter (regress) the processes of linear development (progress). Depending upon the mix of processes that characterize each soil, the pathways of development vary, leading each soil toward different profile characteristics. Third, gradual and steady change in soils might be interrupted by sudden changes in processes triggered not by environmental change but by internal modification of soil-

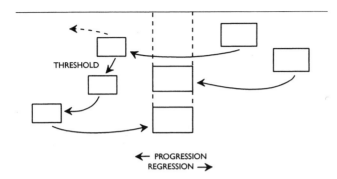

Fig. 1.6. The historical-event model of soil development. The symbols are explained in the caption to figure 1.1.

← PROGRESSION
REGRESSION →

forming mechanisms, internal thresholds in soil development.[15] An example was provided by McFadden and Weldon, who described the steady accumulation of wind-deposited dust in the mountains of southern California, eventually reaching a critical level when, abruptly, the altered texture changed both permeability and water relations and thus the clay concentration in the soil.[16] Fourth, soils may be sensitive to minor differences in initial conditions, which may cause them to move along distinctive developmental pathways, a perspective favored by Phillips in his evaluation of the soils of the North Carolina coastal plain.[17] (However, a champion of the Jenny view might say that different initial conditions simply describe different state factors.)

Ironically, those concerned with the genesis of soil characteristics or general soil profiles have a much greater appreciation of the importance of history than does the traditional student of vegetation dynamics,[18] and yet unlike plant ecology, the science of soil development remains wedded to the adaptation/progress perspective. Vegetation dynamics, in contrast, seems to have been more enriched by the viewpoints of the historical-event model. Perhaps the utility of soils as archives of environmental history prompts this bias. Like the use of pollen and macrofossils for environmental reconstruction, the use of soils for the same purpose requires an assumption that the soil characteristics indeed reflect former environments.

Landforms

Geomorphology, the study of landforms, encompasses a wide range of subjects, but its core might be described as involving the erosional work of streams and of slope processes. Certainly, these two topics were the focus of a dominant figure in geomorphology, William Morris Davis.[19] Davisian geomorphology, moreover, like the classic frameworks in both vegetation dynamics and soil genesis, strongly reflects the viewpoints of Gould's biases (fig. 1.7). First, the peneplain could be seen as the only stable landform configuration, the others being temporary stages, and in its persistence resembled the *adaptational* forms of climax vegetation and equilibrium conditions in soils. Second, the pathway leading toward the peneplain end point describes a *progressive* development of erosional forms, even employing the metaphor of life history employed by vegetation succession—youth, maturity, and old age. Third, the stages of development proceed slowly, *gradually*, particularly compared with the initiation of the erosional pathway by rapid, rejuvenating uplift. Fourth, the initial conditions of the pathway have nothing to do with the character of the subsequent erosional forms; rather, any time that a portion of the earth's crust is uplifted, the erosion that follows adheres to a *deterministic* sequence of stages common to all cycles of erosion, regardless of specific circumstances.

The vision of the developmental pathway progressing toward a state of goodness—conspicuously strong with vegetation, subtle at best with soils—completely disappears with landforms. That geomorphic forms or processes might generate a moral sense in the human mind is indicated by positive attitudes toward mountains and the perfection perceived in the hydrologic cycle.[20] Nonetheless, neither geomorphologists nor nature lovers suggest that virtue rests in the pediment or the peneplain.

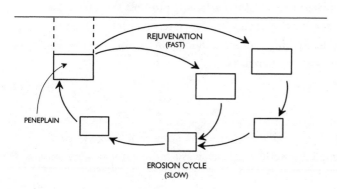

Fig. 1.7. Davisian landform development (adaptation/progress model). Symbols are explained in the caption to figure 1.1.

Thomas R. Vale

Davisian geomorphology has virtually disappeared from the scholarly literature on erosional topography, but elements persist in studies of slopes. For example, the concept of morphogenetic regions—the linking of distinctive erosional processes and forms to areas of particular climate (and sometimes of rock types)—resembles the *adaptational* regionalization of vegetation and soil orders (fig. 1.8).[21] Moreover, models of slope development may conform to the *gradualism* and linear *progression* of the Davisian perspective, as does Oberlander's characterization of scarp retreat in massive sandstone of arid regions.[22]

In scholarship on streams, however, the demise of the Davisian perspective is complete; it has been replaced by traits from the historical-event viewpoint. A classic paper by Womack and Schumm serves as an example.[23] Their study evaluates the recent fluvial history of a watershed, Douglas Creek, in Colorado, where, beginning about the turn of the century and perhaps triggered by livestock grazing, the stream began to erode actively. But unlike the simple cycle of erosion described in the Davisian scheme, Douglas Creek revealed a "complex response": The initial incision increased the sediment load of the stream, which caused deposits of alluvium in a downstream location, which in turn reduced the stream's sediment load and resulted in additional incision. Tributary streams may or may not add to the sediment load of the main stream, and meanders may or may not be cut off, both of which add to the complexity of the response (fig. 1.9). Taken together, this vision of stream behavior confounds single reactions to environmental change (stream response can be either erosion or deposition), denies simple and linear pathways of development (the same locale experiences alternating episodes of erosion and deposition), recognizes the importance of initial conditions as a cause for the variable response (a locale with previous sediment deposition from either a tributary or the main stream may undergo erosion, whereas some other locale—recently eroded or lacking a tributary—may experience deposition), and sees the changes as pulsed rather than gradual (both erosion and deposition occur rapidly, in pulses). Elements of the historical-event model replace those of the adaptation/progress perspective.

Other recent geomorphological work illustrates the dominance of components of the historical-event viewpoint. Rhodes found that characteristics of stream channels, when altered by a flood, may not return to their preflood magnitudes, instead converging upon a different set of stable values[24]—a

Mean annual precipitation (inches)

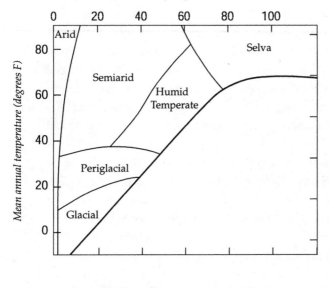

Fig. 1.8. Morphogenetic regions identified for erosional topography. Adapted from D. Ritter, *Process Geomorphology*, 3rd ed. (Dubuque, Iowa: W. C. Brown, 1995).

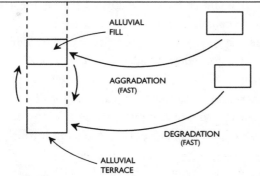

Fig. 1.9. The historical-event interpretation of erosional dynamics on Douglas Creek, Colorado, studied by W. R. Womack and S. A. Schumm ("Terraces of Douglas Creek, Northwestern Colorado"). Symbols are explained in the caption to figure 1.1.

response that suggests a lack of narrow *adaptationism*. Graf described stream behavior counter to rigid *determinism* in a Utah watershed, where different parts of the basin might simultaneously experience erosion or deposition and where the same locale might shift its response from erosion to deposition without being affected by a major outside force.[25] Knox suggested that patterns of pulsed, rather than *gradual*, change characterize the Holocene history of climate, vegetation, and stream response in the upper Midwest (fig. 1.10),[26] and he traced that geomorphic history as responding to climatic episodes,

Thomas R. Vale

which created corresponding periods of high or low flooding without directional, or *progressive*, development through time.[27]

Overall, then, the adaptation/progress perspective, associated with William Morris Davis, has not simply been enriched by the historical-event viewpoint, as seems the case with vegetation dynamics, but seems to have been completely replaced by it in geomorphic work. Elements of adaptation/progress (and even classic Davisian) visions remain, however, as suggested by continuing interest in reconciling apparently conflicting perspectives.[28]

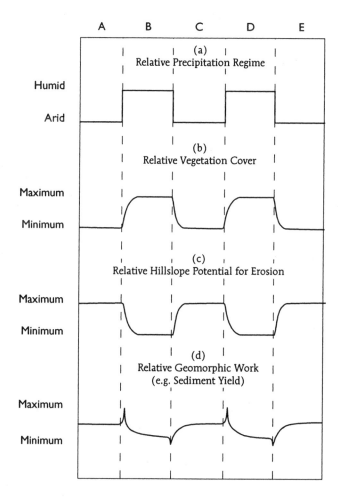

Fig. 1.10. Biogeomorphic-response model of Knox ("Valley Alluviation in Southwestern Wisconsin"). Chronological time progresses on the horizontal scale, and environmental attributes extend on the vertical scale.

Weather and Climate

Research on atmospheric processes and forms, unlike the scholarship in the areas of vegetation dynamics, soil genesis, or erosional landforms, defies analysis in terms of either the dichotomy between the adaptation/progress and historical-event models or the four elements in each of the two perspectives. Part of the difficulty resides in the differing visions that result from the extremes in time scales of interest to those concerned with the atmosphere—from hours to millions of years. But even restricting an evaluation to a particular time frame suggests the use of unique mixes of elements within the models rather than the aggregated models as wholes.

The time scales that cover from thousands to hundreds of thousands of years—such as the Holocene or the Pleistocene period—serve as initial examples (fig. 1.11). First, the atmospheric response to outside forcing, such as variations in energy receipt from the sun, or boundary conditions, such as snow and ice cover at high latitudes, is sufficiently specific to allow the development of circulation models,[29] an accomplishment that assumes *adaptationism*. Second, the recorded variations in climatic conditions, such as the alternating cold and warm eras during the Pleistocene, invite explanation in terms of a recurring cyclical pattern in the causative factors—the same mix of causes induces a *deterministic* single atmospheric response. These two adaptation/progress elements mix, however, with two elements of the historical-event perspective: viewed over many thousands of years, the breaks between major cold or warm periods seem abrupt, rather than *gradual*, and the recurrence of the periods lacks directional, *progressive* patterning.

A shortening of the time scale again suggests that neither aggregated model describes atmospheric behavior; rather, a mix of elements from the two characterize explanations. The phenomenon of El Niño serves as an example. In spite of popular news reporting that portrays a legion of specific and predictable responses in the middle latitudes to an El Niño in the tropical Pacific, much more variable linkages tie the two regions of the world. Namias studied the nine El Niño events from about 1945 to the mid-1980s and found nine different winter-precipitation and six temperature-anomaly patterns over the United States[30]—multiple responses of the atmosphere of the northern middle latitude to El Niño events, an apparent denial of rigid *adaptationism* or *determinism*. On the other hand, the well-documented El Niño events of

Thomas R. Vale

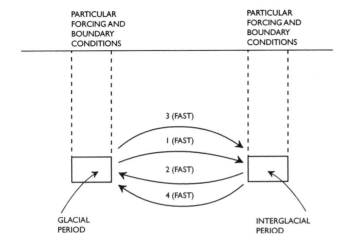

PARTICULAR FORCING AND BOUNDARY CONDITIONS

PARTICULAR FORCING AND BOUNDARY CONDITIONS

3 (FAST)

1 (FAST)

2 (FAST)

4 (FAST)

GLACIAL PERIOD

INTERGLACIAL PERIOD

Fig. 1.11. Interpretation of the dynamics of climate periods of the Pleistocene. Symbols are explained in the caption to figure 1.1.

recent years (that of 1997/98 being particularly accessible as a result of satellite graphics and computer technology) reveal *gradualism* in *progressive* development toward the El Niño peak and a similar linear pattern in its waning.

Nor does a given time scale become linked with any particular expression of one of the four elements. The examples of long time scale noted above describe adherence to adaptationism and determinism, but in a different context (generalized energy-balance models of the atmosphere appropriate for such time scales) Ghil identified multiple configurations of atmospheric circulation reacting to a given set of forcing factors.[31] In contrast, the short-time-scale examples cited were associated with gradual, linear developmental pathways, but even at the scale of decades or years abrupt, nonprogressive changes are often identified.[32]

End points associated with virtuous conditions do not seem to be linked to directional pathways in atmospheric science. El Niño peaks (or their nadirs, La Niña peaks), mature midlatitude cyclones, large thunderstorms or hurricanes, marches of the seasons toward equinoxes, global or hemispheric temperature increases in the final decades of the twentieth century—none of these common linear developments in atmospheric phenomena (linear at least when viewed in the context of an appropriate time scale) suggest progress toward enhanced qualities. In fact, in many of these examples the progression generates worrisome concern more often than welcoming praise. When evaluating the atmosphere, more than in other subjects discussed, we humans seem to see virtue in the status quo rather than in pathways of change.

The study of the atmosphere, then, seems not to be dominated by particular interpretive perspectives that loom central to analyses in vegetation, soils, and landforms. That is, scholarship in weather and climate employs no organizing perspective and, moreover, as a consequence of this lack also does not experience paradigm shifts or even a history of varying interpretations—the sorts that link other subjects to such names as Clements, Jenny, and Davis. Perhaps the interpretation of atmospheric phenomena is driven by first principles, by physics, and thus finds less need for conceptual models.[33]

Reflections

In summary, interpretations in the various parts of physical geography vary in their invocation of the two aggregated models and the four elements that both models contain. Vegetation dynamics continues to reflect strongly its traditions in the adaptation/progress perspective but is enriched by the contrary characteristics inherent in the historical-event view; more than in other subjects of physical geography, moreover, vegetation dynamics embraces a moral sense of betterment in the progressive development of plant covers. Soil genesis more rigidly than the other subjects adheres to the adaptation/progress model, with only hints of the complexities inherent in the contrasting views of the historical-event perspective; virtuous end points to linear development may appear in soils studies only in the preferences of subjects deemed worthy of study. Erosional landform analysis seems to have rejected more completely, compared with the other subdisciplines, an earlier dedication to the adaptation/progress model and embraced much of the viewpoint associated with historical events; no progress toward virtue is a part of landform development. The interpretation of weather and climate relies little on the conceptual models that are so much a part of the thinking in the other subject matters, with elements of the two perspectives employed only as required by the patterns revealed by data; as with landforms, no sense of betterment is associated with the linear development of phenomena in the atmosphere.

Why should such differences exist in the subdisciplines of physical geography? It may be tempting to seek explanation in such concepts as paradigm shifts, ages of disciplines, or lines of argument that tie science to social

Thomas R. Vale

construction of knowledge. Other, more direct interpretations seem more tenable. First, the pace of development, of change, varies in the phenomena of the four subject areas, particularly when each is compared with the time scale of human life experience. Soils (at least many soil characteristics) develop most slowly (centuries or more) and thus may require a conceptual model to organize the trends and events—the adaptation/progress model, with its linear form of development, allows humans to understand the genesis of soil profiles. Vegetation recovering from a disturbance event also unfolds at a slow rate—from many decades to a century or two—and again, the adaptation/progress model helps humans to understand the phenomenon; the more rapid pace of change, compared with that of soils, however, makes vegetation dynamics less firmly wedded to the linear model. In erosional geomorphology the shift from concern with Davisian form to contemporary process shortened, in essence, the time scale of study and in so doing reduced the appeal of, or the need for, the adaptation/progress perspective in favor of the shorter time frame of the historical-event viewpoint; this transformation, nearly complete in scholarship on streams, seems less universal in the study of erosional slopes (the processes of which typically operate slowly compared with those of streams), where both a focus on form and a dependence on the adaptation/progress model persist. In weather and climate the description of change has not been intimately linked to a conceptual perspective, perhaps because the pace of variability in atmospheric phenomena seems much more rapid than in the other subject areas.

A second explanation for the differential appeal of the two models lies in the nature of the processes invoked to explain observations, that is, in the closeness of the subject matters—and the explanatory ties—to basic sciences.[34] As noted earlier, the tight connection between atmospheric process and physics may reduce the appeal of conceptual, organizing models; by contrast, the other areas of physical geography may focus more on end-result forms and thus find such models useful. From this stance, geomorphology nestles much closer to atmospheric process than does vegetation dynamics and, accordingly, depends less on the more rigid conceptual model of the adaptation/progress perspective. Soil genesis, oddly, remains a puzzle. Scientific explanation of causation in soils employs the physical processes familiar to the atmospheric scientist (the intimate scale makes soils more like atmospheric processes than even geomorphology), and yet conceptually soil genesis resembles the explanatory viewpoint of vegetation more than that of its sibling earth science.

Overall, then, what about Gould's assertion that progress, adaptationism, gradualism, and determinism permeate modern science? In physical geography his pronouncement does not seem to be universally appropriate: How strongly these traits appear in vegetation dynamics, soil genesis, erosional geomorphology, and atmospheric processes varies greatly, perhaps prompted by differing paces of change in each subarea of the natural world and by distinctive linkages to basic sciences, that is, physics and chemistry. In Gould's own area of specialty, evolution, the pace of change is extremely slow (or at least the time scale in which changes occur is long, even with punctuated equilibrium; change may be rapid but set within extended periods of stasis), and the explanation narrative is distanced from those in the fundamental sciences—both characteristics may prompt the continuing appeal of the adaptation/progress model, of Gould's four biases.

Gould does not reject progress—either linear development or movement toward betterment—categorically. In the scientific understanding of the universe, in the performing arts, and in the creative arts Gould finds no limits to human development, to progress.[35] (Gould clearly asserts that scientific knowledge grows and thus progresses toward ever-greater knowledge about the world.) His is a viewpoint that readers of Yi-Fu Tuan will find familiar. One ingredient of the "good life" comes from the richness of personal and cultural experience:

> We may see the high moments and achievements, our own and society's, as *steps* to the pleasing present and to an even better future. . . . This is the forward-looking dynamic view, with stress on accumulation and a subjective sense of progress. . . . I have produced a personal museum; and as I walk slowly and appreciatively through its galleries . . . I feel that my own life—and world—has been progressively enlarged. That sense of accumulation—that history is cumulative—comes out of an awareness that if I were a Roman, my stroll would have terminated with the Etruscans and the Greeks.[36]

The "pleasing present" and "sense of accumulation"—such optimism speaks to the virtues of the Enlightenment, and the Enlightenment gives us progress all the way.

Thomas R. Vale

Notes

1. S. J. Gould, *Full House: The Spread of Excellence from Plato to Darwin* (New York: Harmony, 1996).

2. See D. R. Stoddart, "Darwin's Impact on Geography," *Annals of the Association of American Geographers* 56 (1966): 683–98; W. H. Drury and I. C. T. Nisbet, "Inter-relations between Developmental Models in Geomorphology, Plant Ecology, and Animal Ecology," *General Systems* 16 (1971): 57–68; S. J. Gould, "Evolution and the Triumph of Homology, or Why History Matters," *American Scientist* 74 (1986): 60–69; B. A. Kennedy, "Hutton to Horton: Views of Sequence, Progression, and Equilibrium in Geomorphology," *Geomorphology* 5 (1992): 231–50; and W. R. Osterkamp and C. R. Hupp, "The Evolution of Geomorphology, Ecology, and Other Composite Sciences," in *The Scientific Nature of Geomorphology: Proceedings of the Twenty-seventh Binghamton Symposium in Geomorphology*, ed. B. Rhoads and C. Thorn (New York: John Wiley, 1996), 415–41.

3. F. E. Clements, *Plant Succession: An Analysis of the Development of Vegetation,* Carnegie Institute Publication 242 (Washington, D.C.: Carnegie Institute, 1916); see also T. R. Vale, *Plants and People: Vegetation Change in North America* (Washington, D.C.: Association of American Geographers, 1982).

4. J. T. Hack and J. C. Goodlett, *Geomorphology and Forest Ecology of a Mountain Region in the Central Appalachians,* Geological Survey Professional Paper 347 (Washington, D.C.: U.S. Geological Survey, 1960).

5. J. W. Thomas, ed., *Wildlife Habitats in Managed Forests: The Blue Mountains of Oregon and Washington,* Agricultural Handbook No. 553 (Washington, D.C.: U.S. Department of Agriculture, Forest Service, 1979); D. A. Perry, *Forest Ecosystems* (Baltimore: Johns Hopkins UP, 1994); W. C. Smith and J. K. Fischer, *Fire Ecology of the Forest Habitat Types of Northern Idaho,* General Technical Report INT-363 (Ogden, Utah: Rocky Mountain Research Station, 1997).

6. B. McCune and T. F. H. Allen, "Will Similar Forests Develop on Similar Sites?" *Canadian Journal of Botany* 63 (1985): 367–76.

7. F. E. Egler, "Vegetation Science Concepts. I. Initial Floristic Composition, a Factor in Old-field Vegetation Development," *Vegetatio* 14 (1954): 412–17.

8. J. D. Henry and J. M. A. Swan, "Reconstructing Forest History from Live and Dead Plant Material—An Approach to the Study of Forest Succession in Southwest New Hampshire," *Ecology* 55 (1974): 772–83.

9. N. L. Christensen, "Landscape History and Ecological Change," *Journal of Forest History* 33 (1989): 45–54.

10. D. C. West, H. H. Shugart, and D. B. Botkin, *Forest Succession: Concepts and Application* (New York: Springer-Verlag, 1981); S. T. A. Pickett and P. S. White, *The Ecology of Natural Disturbance and Patch Dynamics* (New York: Academic Press, 1985); W. A. Laycock, "Stable States and Thresholds of Range Condition on North

American Rangelands: A Viewpoint," *Journal of Range Management* 44 (1991): 427–33; R. J. Tausch, P. E. Wigand, and J. W. Burkhardt, "Viewpoint: Plant Community Thresholds, Multiple Steady States, and Multiple Successional Pathways: Legacy of the Quaternary?" ibid. 46 (1993): 439–47.

11. P. J. Bartlein, K. H. Anderson, P. M. Anderson, M. E. Edwards, C. J. Mock, R. S. Thompson, R. S. Webb, T. Webb III, and C. Whitlock, "Paleoclimate Simulations for North America over the Past 21,000 Years: Features of the Simulated Climate and Comparisons with Paleoenvironmental Data," *Quaternary Science Reviews* 17 (1998): 549–85.

12. D. B. Botkin, *Discordant Harmonies: A New Ecology for the Twenty-first Century* (New York: Oxford UP, 1990).

13. H. Jenny, *Factors of Soil Formation: A System of Quantitative Pedology* (New York: McGraw-Hill, 1941).

14. D. J. Johnson and D. Watson-Stegner, "Evolution Model of Pedogenesis," *Soil Science* 143 (1987): 349–66.

15. D. R. Muhs, "Intrinsic Thresholds in Soil Systems," *Physical Geography* 5 (1982): 99–110.

16. L. D. McFadden and R. J. Weldon, "Rates and Processes of Soil Development on Quaternary Terraces in Cajon Pass, California," *Geological Society of America Bulletin* 98 (1987): 280–93.

17. J. D. Phillips, "Progressive and Regressive Pedogenesis and Complex Soil Evolution," *Quaternary Research* 40 (1993): 169–76.

18. V. T. Holliday, "The 'State Factor' Approach in Geoarchaeology," in *Factors of Soil Formation: A Fiftieth Anniversary Retrospective*, Special Publication No. 33 (Madison, Wis.: Soil Science Society of America, 1994), 65–85.

19. W. M. Davis, *Geographical Essays* (Boston: Ginn, 1909).

20. M. H. Nicholson, *Mountain Doom and Mountain Glory* (Ithaca: Cornell UP, 1959); Yi-Fu Tuan, *Hydrologic Cycle and the Wisdom of God: A Theme in Geoteleology* (Toronto: U of Toronto P, 1968).

21. C. E. Thorn, "Periglacial Geomorphology: What, Where, When?" in *Periglacial Geomorphology*, ed. J. C. Dixon and A. D. Abrahams (New York: John Wiley, 1992), 1–30; M. Thomas, "Geomorphology from a Tropical Perspective," in *Geomorphology in the Tropics* (New York: John Wiley, 1994), 3–15; C. R. Twidale and Y. Lageat, "Climatic Geomorphology: A Critique," *Progress in Physical Geography* 18 (1994): 319–34.

22. T. Oberlander, "Slope and Pediment Systems," in *Arid Zone Geomorphology* (New York: John Wiley, 1989), 56–84.

23. W. R. Womack and S. A. Schumm, "Terraces of Douglas Creek, Northwestern Colorado: An Example of Episodic Erosion," *Geology* 5 (1977): 72–76.

24. B. L. Rhodes, "Mutual Adjustments between Process and Form in a Desert Mountain Fluvial System," *Annals of the Association of American Geographers* 78 (1988): 271–87.

Thomas R. Vale

25. W. L. Graf, "Spatial Variation of Fluvial Processes in Semi-arid Lands," in *Space and Time in Geomorphology*, ed. Colin Thorn (London: Allen & Unwin, 1982), 193–217.

26. J. C. Knox, "Valley Alluviation in Southwestern Wisconsin," *Annals of the Association of American Geographers* 62 (1972): 401–10.

27. J. C. Knox, "Responses of Floods to Holocene Climatic Change in the Upper Mississippi Valley," *Quaternary Research* 23 (1985): 287–300.

28. S. A. Schumm and R. W. Lichty, "Time, Space, and Causality in Geomorphology," *American Journal of Science* 263 (1965): 110–19; C. E. Thorn, "Time in Geomorphology," in *An Introduction to Theoretical Geomorphology*, ed. Thorn (London: Unwin, 1988), 53–72.

29. COHMAP (Cooperative Holocene Mapping Project) Members, "Climatic Changes of the Last 18,000 Years: Observations and Model Simulations," *Science* 241 (1988): 1043–52; Bartlein et al., "Paleoclimate Simulations for North America over the Past 21,000 Years."

30. J. Namias, reported in M. Ghil, "Predictability of Planetary Flow Regimes: Dynamics and Statistics," in *Toward Understanding Climate Change*, ed. Uwe Radok (Boulder: Westview Press, 1987), 91–147.

31. M. Ghil, "Energy-Balance Models: An Introduction," in *Climatic Variations and Variability: Facts and Theories*, ed. A. Berger (Dordrecht, Holland: D. Reidel, 1981), 461–80.

32. R. A. Kerr, "Research News: Unmasking a Shifty Climate System," *Science* 255 (1992): 1508–10; G. Bond, W. Showers, M. Cheseby, R. Lotti, P. Almasi, P. DeMenocal, P. Priore, H. Cullen, I. Hajdas, and G. Bonani, "A Pervasive Millennial-Scale Cycle in North Atlantic Holocene and Glacial Climates," ibid. 278 (1997): 1257–66.

33. J. Burt, personal communication, Madison, Wisconsin.

34. Osterkamp and Hupp, "Evolution of Geomorphology, Ecology, and Other Composite Sciences."

35. Gould, *Full House*.

36. Yi-Fu Tuan, *Good Life* (Madison: U of Wisconsin P, 1986), 159–60.

Chapter 2

Landscape, Place, and the State of Progress

Kenneth R. Olwig

Briefly, I shall argue that we cannot, by taking thoughtful and deliberative steps, maintain a state of rootedness, whereas a sense of place can indeed be thus achieved and maintained. Rootedness implies being at home in an unselfconscious way. Sense of place, on the other hand, implies a certain distance between self and place that allows the self to appreciate a place.
—Yi-Fu Tuan

The "deliberative steps" that bring the peripatetic mind a sense of place and community are, I would argue, inherently circuitous. The steps of the pilgrim's progress thus take the pilgrim away from the homeplace, but they also bring the pilgrim back to the place of origin, allowing the pilgrim to learn to know it afresh.[1] It is through such peregrination that the liminalities that bound the structures of daily life are transcended and a sense of place and community is generated.[2] A sense of place can help us, in Tuan's words, "forget our separateness and the world's indifference": "Place supports the human need to belong to a meaningful and reasonably stable world, and it does so at different levels of consciousness, from an almost organic sense of identity that is an effect of habituation to a particular routine and locale, to a more conscious awareness of the values of middle-scale places such as neighborhood, city, and landscape, to an intellectual appreciation of the planet earth itself as home."[3]

There is a contradiction, however, between the progress that is made, metaphorically speaking, by the taking of deliberate peripatetic steps and the progress that is made when we march, in step, into a future that obliterates the places of its past.[4] According to Yi-Fu Tuan, writing in the journal *Progress in Geography:* "The pedestrian advances by leaving step after step behind him. . . . This commonplace observation gains interest if we think how radically space-and-time changes when a person is not walking but marching with a band. The marching man still moves, objectively, from A to B; however, in feeling open space displaces the constrained space of linear distance and point locations. Instead of advancing by leaving steps behind the marching man enters space ahead."[5] If we follow this line of thinking, the march of progress takes us away from place. The person marching in time to the beat of a drum does not derive such a sense of place,[6] and people marching in lockstep do not form a community.[7] The creation of such a sense of place, I argue, represents a form of peripatetic progress that is in conflict with the march of progress. The march of progress takes us into an "open," placeless space, a *utopia* (or non-place), and tends to obliterate the places that it leaves behind. In what follows I will critically examine the march of progress and contrast it to a *topian* concept of progress, which does not parade us linearly to the steady beat of its drum but, like the spiral of a harmonic progression, allows us to return to, and regenerate, the places that give us sustenance.

Introduction

This is about the transformation of progress from a term that in the Renaissance still identified the circuitous process by which the place of a human commonwealth was generated to a notion that involved the linear development in time and space of a national landscape scene through a succession of stages. This linear idea of progress simultaneously designated its opposite. Forms of place identity rooted in the past experience of society were thereby designated as being in opposition to the march of progress and, by extension, in its way. The idea of progress hereby implied the necessity of eliminating that which impedes progress. This darker, Faustian side of progress might be termed the *dialectic of modernity.* It is a dialectic that opposes cosmic dreams of a *utopian* future to the *topian* exigencies of present and past. Like the ever receding vanishing point of the lines of perspective on the landscape horizon, however, this utopian idea of progress never seems to fully materialize. This is because constituting the social landscape as a progressively changing, nebulous scene has the effect of dissolving the substantive qualities of place to which the progressive functioning of communities is linked. If it is to have any meaning progress must bring us back to the places and communities from which our social existence derives its substance. This essay, by the same token, is not based upon a linear analysis of progressive stages but seeks, rather, to leap between the poles of a discourse that, although it is seen to have taken form in the Renaissance, is nevertheless consciously present in the narratives of twentieth-century modernism.

Progressive Custom

When the Tudor queen Elizabeth I (1533–1603) made her "progress" through England she made a customary circuit from place to place in which the local constituencies of the country were made manifest to the monarch, and the monarch to the country. She sat upon a throne, called the *state,* and was carried in a stately procession through political landscapes shaped by different legal communities according to mutually acknowledged rights of custom. This was a ritual, seasonal process by which a larger and more abstract notion of England, as the place of a commonwealth, was generated. The countermovement to the queen's progress out into the countryside was the movement of the *members* of the *body* of Parliament to their meeting in the capital, which also progressively reinforced the idea of a larger English society under a higher form of law.[8] The abstract notions of justice expressed in common law were likewise generated through the circuitous progress of circuit court judges through legal realms rooted in local custom. The country of England was a country of nested countries in that each county or shire was also thought of as a country.[9] Each shire in turn was made up of nested communities and places. At the end of Elizabeth's reign this progressive process of commonwealth identity formation was seriously contested, however, by a new notion of progress that sought to appropriate established ideas of community and place within the framework of a centralized, absolutist state.

During the reign of Elizabeth's Stuart successor, King James I (1566–1625), the meaning of progress began to undergo the transformation that led to the modern idea of progress as a linear movement, occurring in space and changing over time, through stages of development.[10] The Stuart court facilitated this change by effectively creating a new means of *envisioning* progress by staging the country as the theater of state in elaborate and expensive court masques. In this theater, located at the London court of Whitehall, the head of state no longer progressed circuitously from place to place but was now positioned upon a fixed elevated throne, while the spectacle of the country and its body politic, seen in the spatial perspective of staged landscape scenery, was paraded before him, progressing from scene to scene before his commanding gaze. In this way a political landscape formerly made up of a multiplicity of places, functioning as the arenas of legally constituted communities, was now appropriated within the space constituted by the scenic illusion of landscape. In teasing apart this transformation of the meaning of progress it becomes apparent that the concept of landscape as scenery may mask as much as it reveals about the nature of progress. The unmasking of landscape raises the question whether the linear conception of progress in space and time is real or illusory? Is it possible that progress, conceived as constituted by the peace and justice of a prosperous commonwealth, is necessarily circuitous?

The Political Landscape

At the time of Elizabeth I, England was a *country* of "countries," much as the states of northern Europe, such as Denmark or Sweden, were *lands* of "lands," each sub-land, or *landscape,* having its own quasi-independent representative legal bodies and corresponding body of law. The monarch's progress was a necessary means of marking the mutual recognition of the legal status of the monarchy, as embodied by the regent, and the quasi-autonomous existence, embedded in their customary law, of the *lands* through which the monarch passed. In this context the Germanic word *land* is synonymous with *country;* Scot*land* is the country, or *land,* of the Scots.[11] The suffixes *shape* and *scape* are cognate. In the countries of northern Europe the landscape was the outcome of the *shaping* of the land as a political entity by the legal bodies that represented the local political community and gave its customs the form of law. According to an ancient Nordic proverb, it was by abiding by law that a land became the abode of a community. The Nordic word for custom was *sædvane*, a word combining "habit" *(vane)* with "seat" *(sæd)*, literally, the habits of custom by which one makes a place the seat of one's dwelling (a sense we still find in phrases like *country seat*). As smaller lands were gradually incorporated into a larger union of lands under a monarchical state, these smaller lands often became known as *landscapes*, thereby suggesting that though they had become part of a larger land, they maintained the shape or quality of a *land*, their "land-shape," or landscape. Jutland thus retained the name of a land, and its particular body of "landscape law," after it became a part of the Danish state under Denmark's monarch.

Even though landscapes were no longer independent lands, they lent their special place identity, as well as their ideas of justice, to the spatially larger land of lands encompassed by the realm of the monarch. As late as the sixteenth century Scandinavian monarchs were still elected to their position, and to officially become monarchs they had to be acclaimed by the representative legal bodies, known as *Ting* ("things" or "moots" [meetings]), of the different lands, or landscapes, encompassed by their realm. The progress of the monarch from land to land, country to country, was thus not entirely unlike that of a modern American presidential candidate, whose circuitous path to power involves campaigning from state to state and who is ultimately elected to be the leader of the American state of states by an electoral college representing the individual (sub)states. Americans maintain the complex place identity that was characteristic of England when the early settlers fled the political transformations brought about by the Stuart court in the early seventeenth century. In differing contexts they can be fiercely local, fiercely American, or even fiercely global.

Court versus Country, Lord versus Landscape

The death of Elizabeth in 1603 created a radically new political situation in England because it brought the Stuart Scottish king James VI to power as James I of England. The resulting personal union of the two kingdoms led James to feel that he had been born to achieve the manifest destiny of uniting Scotland and England as "Britain." He began his reign with a magnificent ceremonial progress from Edinburgh to London and marked the inception of his new London-based court with a series of theater masques with a landscape vision of a Britain united under his gaze. It was necessary to promote the court's vision of a united Britain in this way because Parliament, to James's great chagrin, opposed his plans for unifying the countries of Britain into one state. Such a unification would have weakened the legitimacy of Parliament because its authority was based, as Chief Justice Edward Coke (1552–1634) was fond of pointing out, upon the moral authority of custom.

Custom was the basis for the common law that united England as a legal community and empowered Parliament as the representative of the differing estates of that community. According to Coke, "Custom lies upon the land." There were, in his view, "two pillars" for custom: "common usage" and "time out of mind." It was on the basis of these pillars that customs "are defined as a law or right not written; which, being established by long use and the consent of our ancestors, hath been and is daily practised."[12] Custom literally lay upon the land because custom was (and still is) progressively established through the daily practice of people in the course of shaping the place of their dwelling.[13] The constitution of the local community was etched into the land through this practice, so that the physical environment of the land became a material reflection of the commonwealth of interests that governed it.

The force of custom was progressively renewed through ritual circuits of movement, in place and community contexts ranging from the royal progress through the country as a whole, to the progress of urban parades, to the progress of villagers as they beat the bounds of their farm lands during Rogation week (fig. 2.1). Though ostensibly based upon "time out of mind" precedence, customary law was in fact progressively brought up to date through the reinterpretation of precedence in light of present circumstance. Eric Hobsbawm has explained the logic of this mode of thought by comparing it to a motor: " 'Custom' in traditional societies has the double function of motor and fly-wheel. It does not preclude innovation and change up to a point, though evidently the requirement that it must appear compatible or even identical with precedent imposes substantial limitations on it. What it does is to give any desired change (or resistance to innovation) the sanction of precedent, social continuity and natural law as expressed in history."[14]

According to Hobsbawm, " 'tradition' must be distinguished clearly from 'custom' " because whereas tradition denies change, custom reaffirms it. As he writes:

The object and characteristic of "traditions," including invented ones, is invariance. . . . "Custom" cannot afford to be invariant, because even in "traditional" societies life is not so. Customary or common law still shows this combination of flexibility in substance and formal adherence to precedent. The difference between "tradition" and "custom" in our sense is indeed well illustrated here. "Custom" is what judges do; "tradition" (in this instance invented tradition) is the wig, robe and other formal paraphernalia and ritualized practices surrounding their substantial action. . . . Inventing traditions, it is assumed here, is essentially a process of formalization and ritualization, characterized by reference to the past, if only by imposing repetition.[15]

Custom, in other words, becomes tradition when it is reified, in accordance with the dialectic of modernity, as the *costume* belonging to an antiquated past.[16]

The stalemate between the Stuart court and the country of England, as represented in Parliament, produced a situation in which James was encour-

Fig. 2.1. Rev. Thomas Wakefield and parishioners breaking into Richmond Park, frontispiece to *Two Historical Accounts of the Making of the New Forest and Richmond Park* (London, 1751). The villagers are making the sort of perambulatory "progress" characteristic of the custom of beating the bounds. This would include breaking down fences and walls deemed to be blocking access to common land. I am indebted to Stephen Daniels for this print.

Kenneth R. Olwig

aged to promote a new conception of land, country, and progress that would legitimize his efforts to unite Britain under his rule. Though the situation of Britain was unique in many respects, James's situation resembled that of other European princes. The basic outlines of this conflict, which became know as "court versus country" in Britain, was known in northern Europe as "lord versus landscape," or *Landschaft*, as it is spelled in German. The historian Otto Brunner described these contradictions on the basis of Austrian material in the following terms:

> The *Land* comprised its lord and people, working together in the military and judicial spheres. But in other matters we see the two as opposing parties and negotiating with each other. Here the Estates appear as the "*Land*" in a new sense, counterposed to the prince, and through this opposition they eventually formed the corporate community of the territorial Estates, the *Landschaft*. At this point the old unity of the *Land* threatened to break down into a duality, posing the key question that became crucial beginning in the sixteenth century: who represented the *Land*, the prince or the Estates? If the prince, then the *Landschaft* would become a privileged corporation; if the *Landschaft*, then it would become lord of the *Land*.[17]

James would have been well informed about the situation on the Continent because he was related to the lords of a number of northern European territories through his wife, Queen Anne of Denmark (1574–1619). He had met many of these relatives in Copenhagen on the occasion of his wedding, in 1589, well before he ever set foot in London.

The court of his brother-in-law, the now legendary King Christian IV of Denmark, was then making progress both toward the unification of the kingdoms of Denmark and Norway and toward the ideal of an absolute monarchy, ruling on the basis of a body of rational statutory law founded not upon custom but upon the timeless natural law elucidated by mathematics and science. The progress of Denmark was visible in the great astronomical research institution established by the astronomer and cartographer Tycho Brahe, whom James made a point of visiting. Brahe used his science in the service of astrology, to predict the return of the golden age as a utopian era under the absolute rational governance of godlike leaders. Science, not least the science of surveying, also provided the techniques by which the Danish state created the cartographic and scenic visions of the realms it sought to unite under its power. When James became king of England, he gave his wife, Queen Anne of Denmark, carte blanche to spend astronomical sums to produce court spectacles of the Stuart state's vision of a reborn British golden age under the sun-god-like gaze of James. She, in turn, hired the budding English surveyor-architect Inigo Jones (1573–1652), then in the employ of the Danish king, as the behind-the-scenes wizard charged with creating these scenes.

Landscape as the Scene of State

The dramatic entrance of a new English concept of landscape upon the political scene of the Stuart court occurred in the preface to *The Masque of Blackness*, the first major theatrical production promoted by Queen Anne: "First, for the Scene was drawn a Landtschap . . . the scene behind seemed a vast sea . . . that flowed forth, from the termination or horizon . . . which (being on the level of the state) was drawn, by the lines of perspective, the whole work shooting downwards from the eye. . . . So much for the bodily part, which was of Master Inigo Jones, his design and act."[18]

The Masque of Blackness was performed on Twelfth Night—Epiphany—1605. It signaled not just the birth of a new year but the recent birth of a new century, the birth of a new Stuart dynasty, and the birth of an imagined new *British* golden age. The masque, a forerunner of opera, was written by the coming English poet laureate Ben Jonson (1573–1637) according to a concept conjured up by the queen. The staging and costumes were done by Jones, the future court "surveyor," or architect. Jones introduced here, perhaps for the first time in English history, the use of perspective stage scenery. Through the vehicle of the masque's scenery the country was transformed into something new, an ideal landscape scene framed according to the techniques of Italianate perspective painting as inspired by visions of a reborn classical utopia.[19] These techniques in turn were derived from the principles of surveying and cartography.[20]

The first masque enacted for the Stuart court was the Twelfth Night performance in 1604 of Samuel Daniel's *Vision of the Twelve Goddesses*, which was performed by Queen Anne and her ladies in the Great Hall at Hampton Court. In this masque a central-perspective stage was not used. Rather, the scenery was scattered about the hall, and the actors and audience circulated among them, according to Tudor custom. The plot concerns the visit of twelve

Kenneth R. Olwig

Greek goddesses to the "Western Mount of *mighty Brittany,* the land of civil music and of rest." Anne played the central figure, Pallas, "the glorious patroness of this mighty Monarchy."[21] The *Masque of Blackness* marked the move of the masque to a Whitehall setting, complete with central-perspective stage scenery and a unified story line.

The reclusive (and paranoid) King James did not make a bodily progress through the landscape in *The Masque of Blackness* but sat elevated above the crowd, fixed on his palisaded throne of state, which in turn was located in the court theater of his capital city (fig. 2.2). The illusion of the sea, which is described as appearing to flow forth from the horizon, is created by the lines of perspective, which, "being on the level of the *state,*" converge upon the eye of the monarch. The head of state, from his privileged, elevated position, possessed a commanding view of the "bodily part" of the masque, the land-scape scenery of his realm and the players upon it. There were few seats in the theater, and one's social *standing* was literally marked by how close one *stood* to the privileged position of the monarch. The masque was a form of total theater, displaying dance, song, and drama within the context of a spectacle created by Jones that was so spectacular that it overshadowed the textual narrative. At a signal the players and audience would suddenly merge in a geometric dance, thus breaking down the barrier between stage and public, theater and daily life.

The players included professionals, leading lords and ladies (among them a blackened Queen Anne), and the most beautiful youths of the realm. Jones dressed the players in revealing costumes that celebrated the bodily parts of the players, who in turn represented the body politic. The invisible lines of perspective, on the other hand, focused upon the surveillant head of the head of state, who was figuratively identified with the sun, whose beams illumi-nated the landscape. The court masque thus represented, in the very struc-ture of its theater, the structure of a nation overseen by the comprehensive gaze of an imperial state that controlled the body politic via invisible, hence secret lines of power, symbolized by the ability of the science of perspective to create the illusion of infinite depth in a three-dimensional, bodily space. In the masques the meaning of the *land* in "*land*scape" shifts subtly from the customarily constituted country of a people to the soil, which, like the floor of a stage, provides the foundation for the scene upon which we play out our lives. It is a floor, however, that is blocked out like the grid of a map, or the

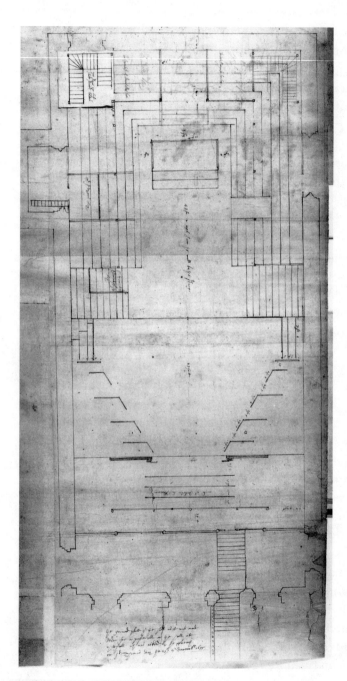

Fig. 2.2. Inigo Jones's sketch of the layout of the hall in Whitehall for the play *Florimène* (1635), plan of the stage and auditorium. The state, or throne, was located in the area marked by a box in the center of the hall opposite the stage, an ideal spot for experiencing the illusion of three-dimensional scenic space. It was not, however, an ideal spot for hearing the text of the play or masque. By permission of The British Library, Lansdowne MS 1171, fols. 5b–6.

Fig. 2.3. Atrium designed by Inigo Jones for the 1632 masque *Albion's Triumph*. This stage scene shows how Jones used perspective stage scenery inspired by classical Roman architecture to envision the new Renaissance architecture he was planning for a reborn Britain. Photograph from the Photographic Survey, Courtauld Institute of Art. Reproduced by permission of the Duke of Devonshire and the Chatsworth Settlement Trustees.

property of an estate. The laws that shape the illusion of bodily, organic space on the stage—the "bodily part"—are not those of historic custom but those of a timeless geometry that marches the eye out into and through an infinite space (fig. 2.3).

The Stuart court's envisioning of the naturally defined geographical body of Britain as the scene upon which the individuals making up the body politic play out their roles creates a space in which one's estate (or status) and hence one's prospects are determined by one's place or position as both observer and participant in the blocked-out spatial framework of the physical British landscape scene. Place ceases to be a landscape or country community de-

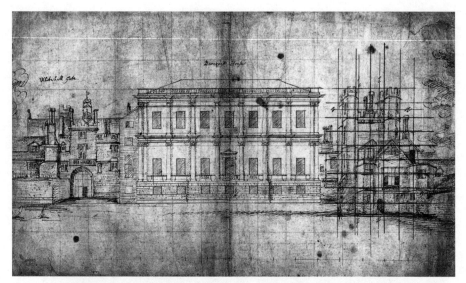

Fig. 2.4. The Banqueting House at Whitehall, by Inigo Jones, drawing from the 1623 masque *Time Vindicated to Himself and to His Honours*. The Banqueting House was built in 1619 to accommodate the productions of masques at court. Note how it contrasts with the traditional Tudor buildings that surround it. The Banqueting House was fitted in 1636 with ceiling paintings by Peter Paul Rubens (1577–1640) showing, among other things, the union of Scotland's and England's crowns under the Stuarts over the dead bodies of the allegorical female figures of Envy and Ignorance. Photograph from the Photographic Survey, Courtauld Institute of Art. Reproduced with the permission of The Conway Library, Courtauld Institute of Art.

fined by an ever evolving body of customary law shaped by practice (fig. 2.4). Instead it is appropriated by the space of the landscape scene, in which individualized bodies occupy a fixed, spatialized location within the coordinates of a larger territorialized framework. The landscape of the theater thereby provides a miniature version of the state of Britain, in which the same lords who flocked to London to position themselves at James's court were expected to manifest their elevated estate at country seats, from which they exercised their power as subregents of the state. This vision of the state was immortalized not long thereafter in the frontispiece to *Leviathan*, by Thomas Hobbes (1588–1679),[22] in which a similarly individualized body politic, with its landscape under the surveillant gaze of the head of state, is given graphic form (fig. 2.5). In Hobbes's vision the individual bodies who play out their

Kenneth R. Olwig

Fig. 2.5. The engraved title page for the 1651 edition of Thomas Hobbes's *Leviathan*. Reproduced by permission of the Cornell University Library, Division of Rare and Manuscript Collections.

roles on its stage have effectively become "decentered" in relation to the larger framework of the theater of state as masque, which is centered, via central-point perspective, upon the gaze of the head of state.[23] They are no longer the authors of their own identity.

Persona, as Hobbes pointed out, "in latine signifies the *disguise,* or *outward appearance* of a man, counterfeited on the Stage; and sometimes more particularly that part of it, which disguiseth the face, as a Mask or Visard: And from the Stage, hath been translated to any Representer of speech and action, as well in Tribunalls, as Theaters."[24]

The idea of the mask is deeply implicated in the way the authority to speak on behalf of the state was constituted in the Renaissance. The world of the court, represented as theatrical *masque,* was a microcosm of the larger theater of the state itself, and those who spoke upon its stage received their persona, and their "authority," from the state as represented in the symbolic form of the theater and its scenery. A person who acted upon the stage of this theater of state might, in Hobbes's words, "represent" either "himself" or "an other." When representing another, "he that owneth his words and actions, is the Author: in which case the Actor acteth by Authority."[25] The theater landscape, like that in the masque, thereby became a means of making visible, as well as audible, the abstract authority of the state, which in this way became the true author of the words and actions of actors upon its stage. Inigo Jones was thus, as Ben Jonson put it, the "wise surveyor! Wiser architect," who would "survey a state."[26]

The Progress of Landscape

The court theater at Whitehall, designed by Jones, envisioned the new conception of the territorial state that the Stuart court was seeking to promote. This state would no longer require the progressive movement of the corporal body of the monarch to hold it together. Instead, it would fix the *head* of state at the *capital* of London, whence the secretive invisible power of the corporate state would permeate and control the landscape and thereby bring harmony and order to an individualized body politic. The court's Lord Chancellor and chief legal council was Francis Bacon, whose support for statutory law based on natural principle made him the natural enemy of Coke and customary law. Bacon proposed an alternative form of government in *The*

New Atlantis, which envisions a utopian autocratic society resembling Plato's Republic ruled by an absolute monarch under the guidance of a college of scientists.[27] The word *sciential* as used at the time is useful because though it literally means "relating to or producing knowledge or science" or "having efficient knowledge,"[28] words like *science* and *knowledge* carried connotations of occult wisdom then. Jones's scenography represented the culmination of a European tradition in which the masque functioned as a forum for the display of the "sciential" power of the state. This notion of science had deep roots in mystical traditions linked to ancient philosophers, astronomers, and astrologers such as Pythagoras, Ptolemy, and Plato and a notion of the state with roots in both Plato and Machiavelli.[29] At this time the state was deliberately enveloped in a secretive aura—the mysteries of state—in part as an attempt to link the state, as the guardian of religion, to the church as the embodiment of the "mystical body" of Christ.[30] Secretive institutions such as the Privy Council and the Star Chamber helped enforce this aura. "Flying machines" and other stage devices, developed with the aid of the most advanced science of the time, helped create a picture of how this "sciential" state would transform the landscape of Britain into that of an ideal golden age, beyond time or history, under the elevated surveillance of the state.

The court masque incorporated elements of the seasonal ritual progress by which the evolving justice of custom was renewed. Seasonal rituals such as those connected with Rogation week, Midsummer, and Yule, were part of a larger yearly framework celebrating the death and rebirth of nature. Elizabeth's progress had been carefully constructed so as to make her person the symbolic vehicle of these natural cyclical powers of renewal. The dramatists of her era likewise incorporated these patterns into their theater. Shakespeare thus built this pattern of change into the structure of plays like *A Midsummer Night's Dream* and *Twelfth Night.* The critic Northrop Frye described this pattern of movement as follows:

> Shakespeare's type of romantic comedy follows a tradition established by Peele and developed by Greene and Lyly, which has affinities with the medieval tradition of the seasonal ritual-play. . . . Thus the action of the comedy begins in a world represented as a normal world, moves into the [natural] green world, goes into a metamorphosis there in which the comic resolution is achieved, and returns to the normal world. . . . Thus Shakespearean comedy illustrates, as clearly as any *mythos* we have, the archetypal function of literature in visualizing the world of

desire, not as an escape from "reality," but as the genuine form of the world that human life tries to imitate.[31]

The progressive changes made in the landscape scenery of the masques followed the general pattern described by Frye, but rather than being made a cyclical seasonal event, the process of vernal renewal was reified as the coming of a revolutionary millennial golden age under a Stuart subvicar of God. The ideal world portrayed in the masque thereby became the symbol of the genuine form of the world against which the inauthenticity of daily life must be measured. Thus, we are told in the conclusion to another Jonson and Jones masque, *The Fortunate Isles, and their Union,* that the unification of Britain signals the coming of a "point of revolution" in which

> There is no sickness nor no old age known
> To man, nor any grief that he dare own.
> There is no hunger there, nor envy of state,
> Nor least ambition in the magistrate.[32]

The Masque of Beauty, the sequel to *The Masque of Blackness,* concludes with these lines:

> May youth and pleasure ever flow;
> But your state, the while,
> Be fixèd as the isle.
> So all that see your beauty's sphere
> May know th' Elysian fields are here.
> Th' Elysian fields are here.
> Elysian fields are here.[33]

The progress toward this millennial rebirth of a mythical ancient Elysian Britain was made manifest by the spectacular changes of three-dimensional scenery architected by Jones. The medium of scenic illusion became the powerful bearer of the Stuart court's message. Furthermore, by causing the eye to follow the lines of perspective out through the landscape and into the infinite potentiality of its distant horizons, this message carried a suggestive subliminal subtext. A classic example of this is found in the masque *Albion's Triumph,* of Twelfth Night, 1632, in which the presence of the monarch prompts the following change of landscape scene:

Kenneth R. Olwig

Here comes the trophy of thy praise,
The monarch of these isles,
The mirror of thy cheerful rays,
And glory of thy smiles:
The virtues and the graces all
Must meet in one when such stars fall.

The King and the masquers dance the main masque, afterward taking his seat by the Queen.

The scene is varied into a landscipt in which was a prospect of the King's palace of Whitehall and part of the city of London seen afar off, and presently the whole heaven opened, and in a bright cloud were seen sitting persons representing Innocency, Justice, Religion, Affection to the Country, and Concord, being all companions of Peace.[34]

According to Stephen Orgel, a leading authority on the English Renaissance masque, this conception of the sciential state listed "among the promised benefits of the new learning the most fabulous wonders of masques: dominion over the seasons, the raising of storms at will, the acceleration of germination and harvest. Every masque is a celebration of this concept of science, a ritual in which the society affirms its wisdom and asserts its control over its world and its destiny."[35] The masque was the product of an idealism that was determined to "purify, reorder, reform, reconceive a whole culture."[36]

The Materialization of the Landscape of Progress

The Stuart court never achieved its goal of creating a united British state under a strong monarch. The second Stuart head of state was literally and symbolically severed from his body just outside the theater in Jones's Banqueting House in 1649. The Glorious Revolution, in 1688, created, however, the basis for a Britain united under a Parliament controlled by a powerful landed oligarchy that no longer needed to fear that unification would throw the balance of power in favor of the monarch. A century after Jonson and Jones staged their landscape vision of a united Britain, Parliament gave its approval to the uniting of the Scottish and English kingdoms. The powerful

---

landed gentry also looked with favor upon the landscape and architectural ideals of the Stuart court. The architects who pioneered the landscape garden parks of Britain and the parks' British Palladian architecture—Richard Boyle, third earl of Burlington (1694–1753), and William Kent (ca. 1685–1784)—were avid students of Jones's designs, and Jones was duly honored with a bust (alongside those of Newton and Locke) in the gallery of British worthies displayed at the gardens designed in part by Kent at Stowe.[37]

For the oligarchy of powerful, commercially and imperially inclined landowners who now controlled Parliament a synonym for *progress* was *improvement*. Improvement was predicated upon the appropriation of the customary use rights of the commoners with the enclosure of their land into bounded and surveyed country estates under their absolute ownership and control. At the same time as these improvers were actively intensifying the agricultural use of the land, they were also busy surrounding their homes with enormous landscape gardens of the sort pioneered by Kent.[38] They now looked out from classical Jonesian/Palladian country seats upon theatrically landscaped scenery in much the same way that King James looked out from his elevated seat upon the British landscape scenery created for the court masque. Like the monarch, they saw this landscape as a symbol of cultural and social progress.[39] The progress signified by these gardens also had a darker side, however: it often required the destruction of whole villages and even the ruination of the "gothic" manorial homes of earlier occupants.[40] The landscape of progress required the destruction of that which was deemed to be its opposite.

Throughout Europe, the landscape park became a symbol of the particular form of progress that Britain was seen to represent. The ascendancy of Britain as the epitome of progress in agriculture and industry was not lost on the remainder of Europe. Charles de Secondat, baron de La Brède et de Montesquieu (1689–1755), was thus an Anglophile who not only promoted British ideals but was so enamored of William Kent's gardens as a symbol of these ideals that he laid out part of his grounds at La Brède in the English style.[41] Perhaps the most encompassing example of the identification between progressive improvement and the landscape garden was in eighteenth-century Anhalt-Dessau, in Germany, where an entire principality was governed on the progressive principles thought to be exemplified by the English-style park constructed on the estate of Prince Franz at Wörlitz.[42]

The Transformation of Progress

To recapitulate, it was in the time of Shakespeare, Jonson, and Jones that the idea of progress began to undergo a transmutation. Instead of simply referring to a progressive, circuitous, movement through space, it began to suggest a linear movement through stages of development and "social progress." As we read in the 1971 *Oxford English Dictionary* under the definition of the verb *progress:*

> **Progress** v. Common in England c. 1590–1670.
> **4.** *fig.* To make progress; to proceed to a further or higher stage, or to further or higher stages continuously; to advance, get on; to develop, increase; usually, to advance to better conditions, to go on or get on well, to improve continuously.
> **1610** B. Jonson *Alch.* II. iii, Nor can this remote matter, sodainly Progresse so from extreme, vunto extreme, As to grow gold, and leape ore all the meanes. **1791** Washington Our country . . is fast progressing in its political importance and social happiness.

This change of meaning can also be seen in the sphere of religion, where the pilgrim in John Bunyan's *Pilgrim's Progress,* from 1678–84, is no longer the sort of pilgrim we know from Geoffrey Chaucer's (ca. 1342–1400) *Canterbury Tales,* whose circuitous progress is to an earthly holy place from which the pilgrim plans to return. Bunyan's inventively named hero and heroine—Christian and Christiana—make their one-way progresses from the earthly "City of Destruction" to the "Celestial Country."[43] At about this same time began the gradual transformation in the meaning of *stage:*

> **Stage:** etym—standing, station, standing place.
> **I. 5.** The platform in a theatre upon which spectacles, plays, etc. are exhibited; esp. a raised platform with its scenery and other apparatus upon which a theatrical performance takes place (1551).
> **IV.** [represents an English development of meaning which seems to have begun about 1600] **10.** A period of a journey through a subject, life, course of action, etc. 1608, Shakespeare (*Per.* iv, iv, 9): "To teach you the stages of our storie."

"All the world," to use Shakespeare's words, had become "a stage."[44] In the hands of Inigo Jones this stage became a landscape scene, and the progress of the state was measured according to the stage-by-stage transformations of that landscape.

The Stages of Progress

The salient feature of the Renaissance stage was the development of perspective scenery. Behind the surface appearance of the scene was the science of mechanics and perspective, which allowed the creation of elaborate stage-works designed to promote the illusion of "bodily" scenic depth based upon the geometries of perspective. The stage thereby came to provide a metaphor for a universe characterized by surface appearance and behind-the-scenes scientific principles that generated that appearance and drove its progress. The power of this metaphor can be seen in the works of Bernard le Bovier de Fontenelle (1657–1757), who was a central figure in promoting what we now regard to be the modern conception of science. Fontenelle wrote in *Conversations on the Plurality of Worlds* (1686):

> I have always thought that nature is very much like an opera house. From where you are at the opera you don't see the stages exactly as they are; they're arranged to give the most pleasing effect from a distance, and the wheels and counter-weights that make everything move are hidden out of sight. You don't worry, either, about how they work. Only some engineer in the pit, perhaps, may be struck by some extraordinary effect and be determined to figure out for himself how it was done. That engineer is like the philosophers. But what makes it harder for the philosophers is that, in the machinery that Nature shows us, the wires are better hidden— so well, in fact, that they've been guessing for a long time at what causes the movements of the universe. . . . Whoever sees nature as it truly is simply sees the backstage area of the theater.[45]

The picture of progress from stage scene to stage scene witnessed in the theater was perceived to be a function of science. This was true of the early-seventeenth-century masques, it was true for Fontenelle, and a century later, in 1795, such ideas informed the influential *Sketch for a Historical Picture of the Progress of the Human Mind,* by Fontenelle's successor at the French Academy of Sciences, Antoine-Nicolas De Condorcet (1743–94).[46] Progress, for Condorcet, was driven by science, but it was manifested in "stages" as if in a theater. Condorcet thus envisioned progress, as his title indicates, as a series of historical pictures through which the linear "march" (his word) of progress took place in a series of stages, from the first stage, when "men are united in tribes," to the contemporary stage, "from Descartes to the foundation of the French Republic," to the tenth and final stage, in which "the future progress of the human mind" was envisioned.[47] Just as in Elizabeth's day, progress still involved movement, but this movement was no longer circuitous; it was now linear and in "march" tempo. Elizabeth's stately progress had been constrained by the need to ever look backward toward custom, whereas Condorcet in his march of progress into the future had his eyes fixed firmly on the horizon ahead:

> How consoling for the philosopher who laments the errors, the crimes, the injustices which still pollute the earth and of which he is often the victim is this view of the human race, emancipated from its shackles, released from the empire of fate and from that of the enemies of its progress, advancing with a firm and sure step along the path of truth, virtue and happiness! It is the contemplation of this *prospect* that rewards him for all his efforts to assist the progress of reason and the defense of liberty. He dares to regard these strivings as part of the eternal chain of human destiny.[48]

Condorcet's own path to the future was obliterated, very shortly after he wrote these words, by the marching soldiers of the revolution that he helped set in motion, but this did not stop those responsible for his death from soon canonizing him as a state hero who had pointed the way to the glorious future of the French nation-state.

Revolutionary Progress

Condorcet paid with his life for a notion of progress that required the destruction of that which was deemed to belong to the past to achieve its goals. A quintessential expression of this destructive dialectic of modernity is Karl Marx's reference in the *Communist Manifesto* to the revolutionary role of the bourgeoisie:

> Constant revolutionizing of production, uninterrupted disturbance of all social conditions, everlasting uncertainty and agitation distinguish the bourgeois epoch from all earlier ones. All fixed, fast-frozen relations, with their train of ancient and venerable prejudices and opinions, are swept away, all newly-formed ones become

antiquated before they can ossify. All that is solid melts into air, all that is holy is profaned, and man is at last compelled to face with sober senses, his real conditions of life, and his relations with his kind.[49]

Marx does not mourn this meltdown of culture and place; rather, he saw it as the necessary precondition for the establishment of a new stage of development. This stage, like the dream of a revolutionary return to the golden age in the Stuart masques, involved the effective end of history and the beginning of a new (Communist) millennium. In order to make way for the new, one had to deny the authenticity of the old and the historic, hence "the Communist revolution" necessarily "involves the most radical rupture with traditional ideas."[50] All that appeared to be solid and traditional must therefore be revealed to be an inauthentic and unnatural ideological construction concealing the true relations of exploitation under the appearance of a palpable permanency. The exposure of this concealed inauthenticity thus paved the way for a timeless utopia that denied being a utopia because it was based upon the rationality of a scientific socialism, which in the sphere of social relations was comparable to the rationality of Darwin's biological theory of evolution.[51] What was deceptive about the dialectic of modernity as presented by Marx in the *Manifesto* was that it required the tacit presupposition of a natural and genuine utopia against which the inauthenticity of the historically constituted could be measured. For Marx this ideal measure was provided principally by the futurism of a communist utopia consciously constructed according to the natural rational precepts of scientific socialism. Marx and (particularly) Engels, however, also saw communism as involving an end of history as hitherto known and a return, in effect, to the supposedly classless conditions that existed prior to this history:

> The whole history of mankind (since the dissolution of primitive tribal society, holding land in common ownership) has been a history of class struggles . . . the history of these class struggles forms a series of evolution in which, now-a-days, a stage has been reached where the exploited and oppressed class—the proletariat— cannot attain its emancipation from the sway of the exploiting and ruling class— the bourgeoisie—without, at the same time, and once and for all, emancipating society at large from all exploitation, oppression, class-distinctions and class struggles.[52]

The dialectic of modernity as articulated by Marx and Engels has its own particular communist content, but this dialectic owes its structure and power, I would argue, to the fact that it implicitly draws upon powerful Western utopian visions, ranging from that of Plato's *Republic* to that of the Judeo-Christian new Jerusalem or Francis Bacon's scientific *New Atlantis*. In these utopias it is the ideal of a future state organized according to timeless rational principles that provides the standard against which the historically constituted present is measured and found wanting. This ideal measure, however, can also be provided by an imagined natural Edenic utopia of a past golden age—Engels's theorizing about primitive communism belongs to this category. Such *pre-* and *post*historical utopian ideals belong, as the word *utopia* indicates, to *no place*, as well as to *no time*. They provide the setting, beyond time and place, wherein mankind can emancipate itself from history and reconstruct its full authentic potential within a society in which, for Marxists, "in place of the old bourgeois society, with its classes and class antagonisms, we shall have an association, in which the free development of each is the condition for the free development of all."[53] If the measure of the authentic is *utopian*, then historically constituted places are inauthentic, and they must be displaced if utopia is to be achieved.

The Non-place of Modernity

For Marx the necessary prerequisite for building the socialist edifice was the revelation that the apparent substantiality of our historically constituted society was in fact a vaporous ideological construct. Only when this society had passed through the cleansing crucible of modernity and was thereby "melted into air" could a space be cleared for the construction of the brave new world of the communist state. This dialectic of modernity, by which progress toward a better future requires the destruction of the place of the past, is by no means peculiar to Marx. It is, as Marshall Berman has shown, foundational to modernism.[54] This dialectic is well illustrated, for example, in the work of the archetypal architect of modernism, Le Corbusier (1887–1965). His modernist visions thus required the simultaneous destruction of large areas of the cities where they were to be realized, for example, Stockholm and Paris. Though Le Corbusier's Parisian proposal was paid for by a capitalist auto-

Fig. 2.6. Palladio's Villa Rotonda, in Vicenza, Italy. Le Corbusier, like Inigo Jones, was inspired both by the architectural example of classical Rome and by the Italian Renaissance architect Palladio. Reprinted from Le Corbusier, *Urbanisme* (Paris: Editions Cres, 1924), 50.

mobile manufacturer,[55] his rhetoric bears a remarkable resemblance to that of Marx and Engels:

> Movement is the law of our existence: nothing ever stands still, for if it does it begins to go backwards and is destroyed, and this is the very definition of life. Therefore we must act, we must advance, we must produce. After a century and a half of miraculous preparation, reason has come into her own in company with science, and science has flung us violently into the machine age. Everything is revolutionized. It seemed as though progress could lead to nothing but universal destruction, but all that crumbled was the old world. Through the debris the new world began to appear boldly. Reason alone, which appeared definitely to dominate everything, might have led us into the deepest despair, but the violent forces of life seem to have thrust us once more into a new adventure. Reason and passion join hands to produce something constructive. . . .
>
> Our world, like a charnel-house, is strewn with the detritus of dead epochs. The great task incumbent on us is that of making a proper environment for our existence, and clearing away from our cities the dead bones that putrefy in them. We must construct cities for to-day.[56]

Le Corbusier's vision involved replacing the places of the past, which he described as "junk," with the space of skyscrapers:

I wish it were possible for the reader, by an effort of imagination, to conceive what such a vertical city would be like; imagine all this junk, which till now has lain spread out over the soil like a dry crust, cleaned off and carted away and replaced by immense clear crystals of glass, rising to a height of over 600 feet. . . . Our city, which has crawled on the ground until now, suddenly rises to its feet in the most natural way, even for the moment going beyond the powers of our imaginations, which have been constrained by age-long habits of thought.[57]

The panoramic landscape prospects seen from these towering edifices would, in Le Corbusier's view, stimulate the imagination, creating visions of a modern future. "In our walks through this maze of streets," Le Corbusier notes, "we are enraptured by their picturesqueness, so redolent of the past." The advent of aerial photography, however, "creates a blow between the eyes," making one aware of the disorganized structure of the city, along with the human misery it caused.[58] To make this elevated vision of the modern "evi-

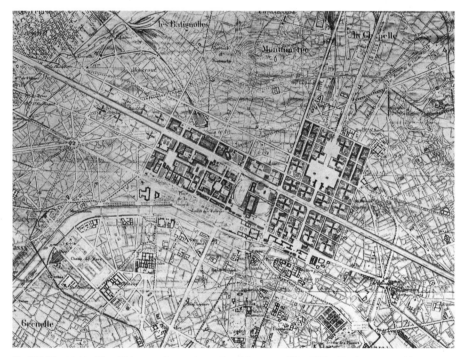

Fig. 2.7. The Voisin Plan. This map shows the scale of the area of "junk" Le Corbusier would have carted away from central Paris, to be replaced with geometric modern structures. Reprinted from Le Corbusier, *Urbanisme*, 272–73.

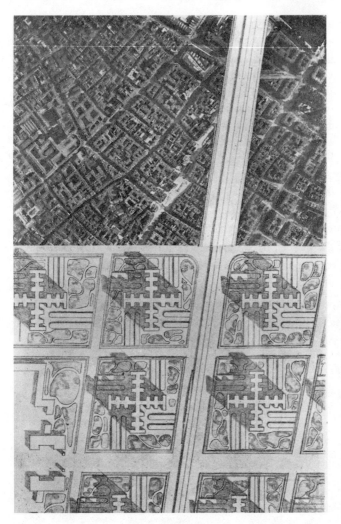

Fig. 2.8. Aerial photograph of part of the area in Paris to be "cleaned off" and a drawing of the structures to be built according to the Voisin Plan. Reprinted from Le Corbusier, *Urbanisme*, 274. The new streets would be straight because, as Le Corbusier put it, "l'homme marche droit parce qu'il a un but, il sait où il va, il a décidé d'aller quelque part et il y marche droit" (man walks in a straight line because he has a goal and knows where he is going; he has made up his mind to reach some particular place and he goes straight to it) (3).

dent *to the eye*," Le Corbusier painted a panorama illustrating his vision of the new Paris. This panorama was displayed in the Pavilion of a journal called *Esprit Nouveau* at the 1925 Paris International Exhibition of Decorative Art. In this panorama one could see "the majestic rhythm of vertical surfaces receding into the distance in a noble perspective and outlining pure forms. From one sky-scraper to another a relationship of voids and solids is established. At their feet the great open spaces are seen. The city is once more based on axes, as is every true architectural creation."[59] Along these axes automobiles raced through the open, linear space.[60] This city was the latest step in a long process of opening up the city, a process that followed "the normal laws of progress."[61]

The Palladian/Jonesian park landscape of Enlightenment Britain underwent a new transformation when it became the preferred setting for modern architecture in the vein of Le Corbusier—a backdrop that provided fine views from the picture windows of apartment towers. According to Le Corbusier, "The City of To-morrow could be set entirely in the midst of green open spaces." One of the mistakes made in New York, in his mind, was that the skyscrapers were not built in "the parks." In Le Corbusier's imagination, "the whole city is a Park. The terraces stretch out over lawns and into groves. . . . Here is the CITY with its crowds living in peace and pure air, where noise is smothered under the foliage of green trees."[62] Much as the "natural" landscape parks created to surround the country seats of the British gentry involved the destruction of villages and their environs, the creation of the lawns,

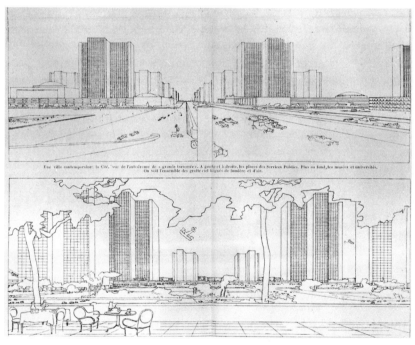

Fig. 2.9. The buildings in Le Corbusier's Voisin Plan are the embodiment of progress as they "march" through the landscape toward infinity. Reprinted from Le Corbusier, *Urbanisme*, 233–34.

sometimes called "green dessert" by environmental planners, surrounding modernist buildings has also involved the destruction of historically constituted places.

Custom versus Modern Progress

Le Corbusier shared the distaste for custom and for democracy expressed by the supporters of Renaissance enlightened despotism. "Town planning," for Le Corbusier, was "the mirror of authority and, it may be, the decisive act of governing." This, to him, suggested the necessity for "a revolution which must be incorporated into the act of governing." "A country as a whole, that is to say the mass of the people as well as their representatives, is not really conversant with the subject [town planning] . . . or with the tremendous possibilities of its emancipating power. They are, on the contrary, entangled in the network of common custom." As envisioned by Le Corbusier, modern town planning would usher in a utopian era not unlike Bacon's *New Atlantis:* "It is undoubtedly true to say that to think in terms of modern town planning is to open every door to harmony and happiness both in the homes and amongst the mass of mankind."[63] To achieve these goals, Le Corbusier concluded, humankind must "make a strong assault on compromise and democratic stagnation."[64]

For Le Corbusier the great prototype of the modern planner was Louis XIV, "the last great town planner in history."[65] Louis's great work had unfortunately become undone "as a result of carelessness, weakness and anarchy, and by the system of 'democratic' responsibilities."[66] For Le Corbusier, however, Louis XIV's work still stood as a beacon for city planners. The last page of his manifesto on urban planning, *The City of Tomorrow,* is thus dedicated to a full-page drawing of Louis XIV "commanding the building of the Invalides" (fig. 2.10).[67]

Conclusion: From Utopianism to Topianism

The idea that custom, misconceived as tradition, is inherently opposed to progress has become so ingrained in Western culture since the Renaissance that this axiom is often taken for granted. Actually, it would be more correct to ascertain that custom stands in the way of the progress of particular social and

Fig. 2.10. "Louis XIV Commanding the Building of the Invalides." Le Corbusier's *Urbanisme* concludes with this drawing. His caption continues: "Homage to a great town planner. This despot conceived immense projects and realized them. Over all the country his noble works still fill us with admiration. He was capable of saying, 'We wish it,' or 'Such is our pleasure.' " Reprinted from Le Corbusier, *Urbanisme,* 285.

political interests and that this helps explain its reduction to static tradition. Progress is a matter of definition. When a community, for example, draws upon legal precedent and custom to argue for the preservation of historical environments in order to protect itself against destruction by a proposed highway or urban renewal project, that community may be seen as hindering the progress of established economic interests. That community, however, might also be seen as unpaving the way for a new appreciation of historical/ environmental values that can radically improve the fabric of urban life.[68] In America the environmental-justice movement is largely community based,[69] and the dramatic reduction of crime in a number of U.S. cities is attributed in part to the reassertion of various forms of community control in the face of the inability of the police to enforce the law.[70] Custom and precedence are still important principles of justice, and communities still reinforce their sense of place identity through all manner of parades, carnivals, pageants, and the like.[71]

From its inception in the Renaissance the modernist dialectic has been embodied in the landscape of a utopian dream of progress as conceived by the architects of modernity (fig. 2.11). To create this utopian landscape it has been necessary to wipe the landscape of *topian* custom from the drawing board. The landscape of custom hindered the progress of particular social interests, such as those behind the (absolutist) central state. It favored, on the other hand, the growth of democracy, as well as conceptions of justice abstracted from the customary law of local communities. Where the modernists once promised to eradicate an inauthentic unnatural past and replace it with an authentic, utopian future, we are now often confronted with certain postmodernist visions of the world in which all seems to be constructed and inauthentic, and progress proceeds from a recognition of this fact.[72] This denial effectively redeploys the modernist dialectic within an intellectual context in which faith in modernism's grand narratives has been lost.[73] Perhaps it is time we moved beyond modernism's *utopianism* and postmodernism's *dystopianism* to a *topianism* that recognizes that human beings, as the creatures of history, consciously and unconsciously create places.

For *topianism* the issue is not whether places are inauthentic constructions, as measured against a utopian ideal, but whether a working community has succeeded in generating a place for "the good life."[74] A word that preserves the sense of a progress that is circuitous is *progression,* as in a *chord*

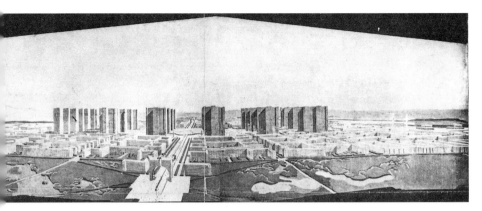

Fig. 2.11. Foldout from Le Corbusier's *Urbanisme* titled "Une ville contemporaine. Vue dioramique de la cité." Reprinted from Le Corbusier, *Urbanisme,* 168a.

progression. Such progressions underlie even complex forms of music. A fully realized chord progression brings us back to the music's harmonic point of origin, leaving us where we started, however enriched. A harmonic progression, as Pythagoras would argue, echoes the sublime but inaudible music of the cosmos at the same time that the melodies and lyrics speak to us of our earthly, placebound experience of hearth and home.[75] *Topianism,* if it is progressive in this sense, is accomplished, on the one hand, through the practices and customs of dwelling and, on the other, through circuitous progressions out into the world that allow us to, in Yi-Fu Tuan's words, "peer beyond the carapaces of place and culture . . . by thus putting a slight distance between us and what we create," thereby helping us to "recognize not only their necessity and power to delude but also their goodness and beauty."[76]

Notes

Epigraph: Yi-Fu Tuan, "Rootedness versus Sense of Place," *Landscape* 24 (1980): 3.

1. In the circuitous steps of the dance we even begin to merge with place because, as Tuan tells us, this brings us into a "homogeneous, nondirected 'presentic' space" (Yi-Fu Tuan, "Space and Place: Humanistic Perspective," *Progress in Geography* 6 [1974]: 226).

2. Victor W. Turner, *The Ritual Process: Structure and Anti-Structure* (New York:

Aldine, 1969); idem, *Dramas, Fields, and Metaphors: Symbolic Action in Human Society* (Ithaca: Cornell UP, 1974).

3. Yi-Fu Tuan, "Place and Culture: Analeptic for Individuality and the World's Indifference," in *Mapping American Culture*, ed. Wayne Franklin and Michael Steiner (Iowa City: University of Iowa P, 1992), 44.

4. The word *peripatetic*, which comes from Greek via Latin, means "given to walking about" or "around," with particular reference to the custom of Aristotle, who taught while walking and discoursing in a place for walking in the Lyceum at Athens (*The Compact Edition of the Oxford English Dictionary*, 1971, s.v. "peripatetic"). Aristotle used this technique to help create the sense of place of an academic community.

5. Tuan, "Space and Place," 225.

6. I distinguish between place and community because even though community often is identified with a particular area in space, it need not be coequal with such an area, and this area may not constitute the place with which the community identifies. Place identity must be socially constituted; it is not determined by the coordinates on a map or by physical conditions. A religious community may be spread over the globe. Such a community, however, may also identify with a particular place (e.g., Rome or Jerusalem) or succession of places (e.g., Jerusalem, Rome). For a useful discussion of the constitution of place see ibid.; Yi-Fu Tuan, *Space and Place: The Perspective of Experience* (Minneapolis: U of Minnesota P, 1977); and idem, "Rootedness versus Sense of Place."

7. By *community* I mean the sort of diverse communality described in the work of Victor Turner, and not regimented unities. The difference can be illustrated by differences in movement and rhythm. Turner focused on the importance of pilgrimage for the development of a sense of "communitas" and, by extension, place identity (Turner, *Ritual Process*; idem, *Dramas, Fields, and Metaphors*). Pilgrims, according to Turner, tend to emphasize the communality of the endeavor by wearing ordinary clothing and using circuitous and pedestrian means of transportation, particularly their feet. The pilgrims' path is circular, and pilgrims are not noted for marching to their destination in lockstep. As Henry David Thoreau was fond of pointing out, pilgrims did not generally stride purposively to their destination but took their time and "sauntered" (Henry David Thoreau, "Walking," in *Nature/Walking*, ed. John Elder [Boston: Beacon, 1991], 71–122). Armies tend to dress in *uniforms*, which emphasize their sense of uniformity, and they move as linearly and uniformly, in march step, as they dress (bringing to mind Emile Durkheim's distinction between "mechanical" and "organic" solidarity). Armies may thus use the march to foster a sense of solidarity among the ranks, but it is a unity in which communality amidst diversity is suppressed—officers tend not to mix with common soldiers.

8. When it was discovered in 1604 that Guy Fawkes had placed a large store of gunpowder under the House of Parliament just before the king was to address it, King James I exclaimed to the assembled M.P.'s that "these wretches thought to

haue blowen vp in a manner the whole world of this Island, euery man being now commen vp here" ("A Speach in the Parliament Hovse, As Neere The Very Words as Covld Be Gathered At The Instant [1605]," in *The Political Works of James I*, ed. Charles Howard McIlwain [Cambridge: Harvard UP, 1918], 286). Thus, the Parliamentarians did not just represent their particular "countries"; they also represented the sum of these countries, the more abstract "community of the English realm" (*communitas regni Angliae*). It was thereby the Parliamentarians who, together with the king, constituted this commonwealth, and with it the *country* of England (Otto Brunner, *Land and Lordship: Structures of Governance in Medieval Austria* [1965; reprint, Philadelphia: U of Pennsylvania P, 1992], 349). The word *country* could thus be used on a variety of levels. On one level it could be identified with the county, and on another with the entire English *commonwealth,* a country founded on law and united by compact or tacit agreement of the people for the common good (commonweal) (*Webster's New Collegiate Dictionary*, 7th ed., s.v. "commonwealth"). The ability of Englishmen to simultaneously identify with the notion of country in a variety of nested spatial realms is illustrated by the words of an M.P. who in 1628 proclaimed: "I speak . . . not for myself, that's too narrow. . . . It is not for the country for which I serve. It is not for us all and the country which we represent, but for the ancient glory of the ancient laws of England" (quoted in Clive Holmes, "The County Community in Stuart Historiography," *Journal of British Studies* 19 [1979–80]: 70).

9. Kenneth Robert Olwig, "Recovering the Substantive Nature of Landscape," *Annals of the Association of American Geographers* 86, no. 4 (1996): 630–53.

10. *Compact Edition of the Oxford English Dictionary,* s.v. "progress." The classic history-of-idea study of progress is J. B. Bury, *The Idea of Progress: An Inquiry into Its Origin and Growth* (New York: Dover, 1955).

11. The Romance languages had similar words carrying similar connotations of legal community and territory. Queen Anne of Denmark's tutor in Italian, John Florio (1553–1625), thus gave the following definition of *paése* in his famed 1611 Italian-English dictionary: "the countrie. Also a countrie, a Land, a region, a province" (*Queen Anna's New World of Words, or Dictionairie of the Italian and English tongues* [London: Edw. Blount & William Barret, 1611], s.v. "paése").

12. Edward Coke, quoted in E. P. Thompson, *Customs in Common* (London: Penguin, 1993), 97, 128, 129.

13. E. P. Thompson, *Whigs and Hunters: The Origin of the Black Act* (New York: Pantheon, 1975), 102; Pierre Bourdieu, *Outline of a Theory of Practice* (Cambridge: Cambridge UP, 1977), 72–95.

14. Eric Hobsbawm, introduction to *The Invention of Tradition,* ed. Terrence Ranger and Eric Hobsbawm (Cambridge: Cambridge UP, 1983), 2.

15. Ibid., 2–3.

16. *Compact Edition of the Oxford English Dictionary,* s.v. "costume."

17. Brunner, *Land and Lordship,* 341. For an English-language historical study that discusses the empowerment of the *Landschaft* in the duchy of Württemberg during

the sixteenth century see David Warren Sabean, *Power in the Blood: Popular Culture and Village Discourse in Early Modern Germany* (Cambridge: Cambridge UP, 1984), 13.

18. Ben Jonson, *The Complete Masques*, ed. Stephen Orgel (New Haven: Yale UP, 1969), 48.

19. Whereas northern European landscape art appears, as the name suggests, to have been largely concerned with representation of the qualities, however idealized, of landscape territories (or countries) as constituted by customary law, Italianate perspectivistic representation of rural scenes was apparently largely concerned with the representation of utopian bucolic ideals found in classical literature (see my discussion of this issue in Olwig, "Recovering the Substantive Nature of Landscape"). The earliest recorded use of the term *Landschaft* to designate the background for a painting dates from 1490 (Walter S. Gibson, *"Mirror of the Earth": The World Landscape in Sixteenth-Century Flemish Painting* [Princeton: Princeton UP, 1989], introduction), and the French equivalent, *paysage*, dates first from 1549, well after the origins of Italianate perspective painting but at the time of the emerging contradictions between *Landschaft* and lord, Protestant and Roman Catholic. *Paysage* appends the suffix *age* to *pays* in much the same way as *schaft* is appended to *Land* in the Germanic languages, or *ship* to *town* in English. *Pays* carried essentially the same connotations of a real community and people as *country* and *land.* The equivalent Italian terms, *paése* and *paesàggio*, emerged about the same time and carried the same meaning (for *pays* and *paysage*, see Ernst Gamillscheg, *Etymologisches Wörterbuch der Französischen Sprache* [Heidelberg: Winter, 1969], and Paul Robert, *Dictionnaire alphabétique et analogique de la langue française* [Paris: Le Robert, 1980]; and for *paése* and *paesàggio*, see Carlo Battisti and Giovanni Alessio, *Dizionario etimologico italiano* [Florence: Instituto Di Glottologia, G. Barbèra, 1975]).

20. See my review of the literature on this subject in Olwig, "Recovering the Substantive Nature of Landscape."

21. Kristin Rygg, *Masqued Mysteries Unmasked: Pythagoreanism and Early Modern North European Music Theatre* (Trondheim, Norway: University in Trondheim, Department of Musicology, 1996).

22. Thomas Hobbes, *Leviathan*, ed. Richard Tuck (Cambridge: Cambridge UP, 1991); originally published as *Leviathan, or The Matter, Forme, & Power of a Commonwealth* ECCLESIASTICALL AND CIVILL (1651).

23. On landscape scenery and the "decentering" of the individual see Kenneth R. Olwig, "Landscapes as a Contested Topos of Place, Community, and Self," in *Textures of Place*, ed. Steven Hoelscher, Paul Adams, and Karen Till (Minneapolis: U of Minnesota P, 2001), 95–119.

24. Hobbes, *Leviathan*, 112, emphasis in the original.

25. Ibid.

26. Ben Jonson, "An Expostulation with Inigo Jones" (1631), in *The Oxford Authors: Ben Jonson*, ed. Ian Donaldson (Oxford: Oxford UP, 1985), 464.

Kenneth R. Olwig

27. Francis Bacon, *New Atlantis* (1627), ed. Alfred B. Gough (Oxford: Clarendon Press, 1915).

28. *Webster's New International Dictionary,* 3rd ed., s.v. "sciential."

29. Frances Yates traces Jones's use of mathematics and mechanics in constructing his stage effects back to the work of such Elizabethan scientists cum mystics and cum magicians as the mathematician John Dee (1527–1608) and the physician Robert Fludd (1574–1637). According to Yates, "In the courts of Europe such shows were prestige symbols, affirmations of the greatness and wealth of the monarchs who could afford them. There was also undoubtedly still lingering some idea that a kind of magic was worked by such spectacles in aid of the monarchs whom they extolled" (*Theater of the World* [London: RKP, 1969], 85–86). The linking of the sciential and the magical and occult was, I would venture, more than "lingering."

30. Ernst H. Kantorowicz, *The King's Two Bodies: A Study in Mediaeval Political Theology* (Princeton: Princeton UP, 1957), 7–23, 193–232; Leonard Barkan, *Nature's Work of Art: The Human Body as Image of the World* (New Haven: Yale UP, 1975), 61–115.

31. Northrop Frye, *Anatomy of Criticism: Four Essays* (Princeton, Princeton UP, 1971), 182–84.

32. Jonson, *Complete Masques,* 447, lines 344–47.

33. Ibid., 74, lines 340–46. Kristen Rygg argues that the echo "is clearly coming *from* Elysium; the voice is a voice from the heavenly spheres and speaks of a divine presence, called forth in the end by the dancing of the worlds' soul" (*Masqued Mysteries Unmasked,* 310). This would make sense given the theme of reflection found in the reference in the *Masque of Blackness* to the Pythagorean message of the moon (which Rygg does not discuss) seen in the water of the Nile.

34. Inigo Jones and Aurelian Townshend, *Albion's Triumph,* lines 330–43, in *Inigo Jones: The Theater of the Stuart Court,* ed. Stephen Orgel and Roy Strong, vol. 2 (Berkeley: U of California P, 1973), 457.

35. Stephen Orgel, *The Illusion of Power: Political Theater in the English Renaissance* (Berkeley: U of California P, 1975), 55.

36. Ibid., 87.

37. George Clarke, "Grecian Taste and Gothic Virtue: Lord Cobham's Gardening Programme and Its Iconography," *Apollo* 97 (May–June 1973): 569.

38. John Barrell, *The Idea of Landscape and the Sense of Place* (Cambridge: Cambridge UP, 1972), 64–97; Raymond Williams, *The Country and the City* (New York: Oxford UP, 1973), 60–107.

39. The term *theater* was applied to areas in gardens in the Palladian style of Kent, and it was commonly applied to landscape gardens by subsequent writers, as in this passage by Horace Walpole about Kent: "Thus, selecting favourite objects, and veiling deformities by screens of plantation; sometimes allowing the rudest waste to add its foil to the richest theatre, he realized the compositions of the greatest

masters in painting." Elsewhere Walpole wrote that "prospect, animated prospect, is the theatre that will always be the most frequented" (Horace Walpole, "The History of the Modern Taste in Gardening" [1782], in *Horace Walpole: Gardenist—An Edition of Walpole's The History of the Modern Taste in Gardening with an Estimate of Walpole's Contribution to Landscape Architecture,* ed. Isabel Wakelin Urban Chase [Princeton: Princeton UP, 1943], 26, 34). On the role of theater and the ideas of Jones in British garden design of this period see also John Dixon Hunt, *Gardens and the Picturesque: Studies in the History of Landscape Architecture* (Cambridge: MIT Press, 1992), 47–102.

40. Nigel Everett, *The Tory View of Landscape* (New Haven: Yale UP, 1994).

41. C. P. Courtney, *Montesquieu and Burke* (Oxford: Basil Blackwell, 1963), 4.

42. Jesper Brandt, "Afgrunden mellem landskabsøkologi og landskabsplanlægning," in *Sådan ligger landet,* ed. Annelise Bramsnæs et al. (Copenhagen: Dansk Byplanlaboratorium, 1987), 127–40.

43. John Bunyan, *The Pilgrim's Progress* (1678–84; reprint, London: Collins, 1953), 183.

44. Shakespeare, *As You Like It,* 2.7.139.

45. Bernard le Bovier de Fontenelle, *Conversations on the Plurality of Worlds* (1686; reprint, Berkeley: U of California P, 1990), 12. Note that Fontenelle uses the plural in referring to the different "stages" presented to the spectator at the opera. The successive scenes presented to the viewer were thought of at the time as changing *stages* rather than, as today, as changing *scenes* upon a single stage.

46. Antoine-Nicolas De Condorcet, *Sketch for a Historical Picture of the Progress of the Human Mind,* trans. June Barraclough (London: Weidenfeld & Nicolson, 1955), originally published in 1795. I would like to thank Tom Are Trippestad for bringing the relevant work of Fontenelle and Condorcet to my attention.

47. Ibid., v–vi, 9.

48. Ibid., 201, emphasis mine.

49. Karl Marx, *Communist Manifesto,* with a preface by Frederick Engels (1848; reprint, Chicago: Henry Regnery, 1969), 20.

50. Ibid., 54.

51. Ibid., 58–78, 8.

52. Frederick Engels, preface to ibid., 7–8.

53. Marx, *Communist Manifesto,* 57.

54. Marshall Berman, *All That Is Solid Melts into Air: The Experience of Modernity* (New York: Simon & Schuster, 1982).

55. Le Corbusier's project to raze a large section of central Paris was named the "Voisin Plan," after the automobile manufacturer who paid for Le Corbusier's exhibition at the pavilion of the *Esprit Nouveau* (an international journal of contemporary activities) at the 1925 Paris International Exhibition of Decorative Arts. Le Corbusier felt that since "the motor has killed the great city, the motor must save the great city" (Le Corbusier, *The City of Tomorrow,* 3rd ed., trans. Frederick Etchells

[London: Architectural Press, 1971], 275; the original French edition appeared in 1924). In the preface to the 1947 edition of *The City of Tomorrow*, written after the orgy of urban destruction resulting from the Second World War had made demolishing historic urban spaces unfashionable, Le Corbusier recanted on his earlier radical views on this subject. He clearly regretted that the Voisin Plan was to be included in the new edition of this historic book (the preface is dated 1945).

56. Ibid., 243–44.

57. Ibid., 281. Le Corbusier was emphatic about the need to make a clean slate out of the existing, historically constituted city. As he put it, "How to create a zone free for development is the second problem of town planning. Therefore my settled opinion, which is quite a dispassionate one, is that the centres of our great cities must be pulled down and rebuilt" (98). "Statistics," he added later, "show us that business is conducted in the centre. This means that wide avenues must be driven through the centres of our towns. *Therefore the existing centres must come down.* To save itself, every great city must rebuild its centre" (116, emphasis in the original). "We must build," he wrote in another place, "*in the open:* both within the city and around it" (176, emphasis in the original). If this was not absolutely clear, he made sure the reader got the message, writing in capitals for emphasis: "WE MUST BUILD ON A CLEAR SITE. *The city of to-day is dying because it is not constructed geometrically. To build on a clear site is to replace the "accidental" lay-out of the ground, the only one that exists to-day, by the formal lay-out. Otherwise nothing can save us. And the consequence of geometrical plans is Repetition and Mass-production*" (220, emphasis in the original).

58. Ibid., 284.

59. Ibid., 282.

60. According to Le Corbusier, "A motor-car which is achieved by mass production is a masterpiece of comfort, precision, balance and good taste" (ibid., 176).

61. Ibid., 283.

62. Ibid., 82, 177. These parks, in Le Corbusier's opinion, could be "of the formal French kind or in the undulating English manner, and could be combined with purely geometrical architecture" (236), but in practice the English has dominated in modernist architecture.

63. Ibid., viii.

64. Ibid., 139.

65. Ibid., 152. "We behold with enthusiasm the noble plan of Babylon and we pay homage to the clear mind of Louis XIV," wrote Le Corbusier; "we take his age as a landmark and consider the *Grand Roy* the first Western town planner since the Romans" (45). Le Corbusier considered Louis XIV's Place Vendôme "one of the purist jewels in the world's treasury" (152). He viewed Louis XIV as a monarch who not only had absolute power but also had "a conception well thought out and clearly presented." It was therefore "useless," according to Le Corbusier, "to say that everything is possible to an absolute monarch; the same thing might be said of Ministers and their Departments, for they are, potentially at least, absolute monarchs (even if

their general slackness prevents their being so in fact)" (151–52). The ideal combination appeared to be an architect of state, like Louis XIV, who had both absolute power and a clear conception of his goal.

66. Ibid., 14.

67. Ibid., 302.

68. J. Douglas Porteous, *Planned to Death: The Annihilation of a Place called Howdendyke* (Manchester: Manchester UP, 1989).

69. Giovanna Di Chiro, "Nature as Community: The Convergence of Environment and Social Justice," in *Uncommon Ground: Towards Reinventing Nature*, ed. William Cronon (New York: W. W. Norton, 1995), 298–320; David Harvey, *Justice, Nature, and the Geography of Difference* (Oxford: Blackwell, 1996), 366–402.

70. Jeffrey Rosen, "The Social Police: Following the law, because you'd be too embarrassed not to," *New Yorker*, 20 and 27 October 1997, 170–81.

71. Turner, *Ritual Process*; idem, *Dramas, Fields, and Metaphors*; Ann-Kristin Ekman, *Community, Carnival, and Campaign: Expressions of Belonging in a Swedish Region* (Stockholm: University of Stockholm, Department of Social Anthropology, 1991).

72. David Harvey, *The Condition of Postmodernity: An Enquiry into the Origins of Cultural Change* (Oxford: Blackwell, 1990), 113–18, 336–37.

73. Jean-François Lyotard, *The Postmodern Condition: A Report on Knowledge* (Manchester: Manchester UP, 1979).

74. On this life see Yi-Fu Tuan, *The Good Life* (Madison: U of Wisconsin P, 1986).

75. Kenneth R. Olwig, "Harmony, 'Quintessence,' and Children's Acquisition of Concern for the 'Natural Environment,'" *Children's Environments* 10, no. 1 (1993): 60–71; Yi-Fu Tuan, *Cosmos and Hearth: A Cosmopolite's Viewpoint* (Minneapolis: U of Minnesota P, 1996). For a thoroughly Teutonic analysis of the social meaning of *harmony* see Leo Spitzer, *Classical and Christian Ideas of World Harmony* (Baltimore: Johns Hopkins P, 1963).

76. Tuan, "Place and Culture," 46.

Kenneth R. Olwig

Chapter 3

The Disenchanted Future

David Lowenthal

Assumptions of progress that have long been deeply embedded in Enlighten-ment perspectives are today much eroded. Confidence in ever improving prospects has in large measure given way to sour surmise that things used to be better than they now are and that the future is likely to be worse. Nostalgia for times past is reinforced by pessimism about what lies ahead. Moreover, much that used to be regarded as progress is nowadays critically scrutinized in the light of diverse realms of existence, diverse criteria of appraisal, and diverse judgments about good and evil. The very title of a major recent over-view, *Progress: Fact or Illusion?* attests these doubts.[1] And the book's troubled contributors echo the assumptions voiced in a Pennsylvania woman's road directions: "Drive straight up the way and you'll hit Progress Road, then just go till you can't anymore. Progress dead-ends at the prison."[2]

Dismay about the future stems in part from its growing uncertainty. The more complex our present world becomes, the less we feel we can rely on the regularities of natural history and the harder it is to foretell the outcomes of our own actions. Unease stems from too much as well as too little knowledge. Enhanced demands of forecasting exacerbate doubts about our ability to cope with innovation, both planned and unintended. Environmental alarms and the pace of unloved change have put paid to utopian visions. In cosmic, genetic, and ecological affairs alike, the presaged future seems more a fear-some apocalyptic planet than a confident Brave New World.

These pervasive anxieties are quite recent. For most of the nineteenth and twentieth centuries the future was a bright and shining presence. Scientific discoveries, social engineering, and the retreat of old-guard tradition engendered cornucopian forecasts. The advances of technology, the visions of architects, and the dreams of science fiction made scenes of progress familiar. Jules Verne's *Five Weeks in a Balloon* (1863) inaugurated countless vistas that peaked with Edward Bellamy's *Looking Backward* (1888) and H. G. Wells's *When the Sleeper Wakes* (1899). That the conquest of nature had enormously enriched the globe was conventional wisdom, and continued scientific progress presaged ever greater advances. Essential to general well-being, the cumulative reshaping of the globe became the normative mode of Western understanding.

Optimism about the benign effects of science was accompanied by faith that future discoveries would reveal the final secrets of nature. Confidence in a better future "seemed to become irresistible in the eighteenth and nineteenth centuries," in one authoritative summary, "and even appeared to move serenely into the twentieth."[3] Turn-of-the-century faith in progress largely survived the First World War, the Great Depression, and even the Second World War.

Architectural previsions were mostly upbeat *villes radieuses*. Planners saw the future as "a period style, a neo-gothic of the Machine Age, as revealed in the Art-Deco skyscrapers of New York in the twenties." In Reyner Banham's depiction, the archetype was "a city of gleaming, tightly clustered towers, with helicopters fluttering about their heads and monorails snaking around their feet; all enclosed . . . under a vast transparent dome."[4] Life in that future, wrote the sardonic Olaf Stapledon, would be "unmitigated bliss."[5]

Neotechnic bliss succumbed to postwar ecological fears and primitivist nostalgia, exemplified in the Luddite flower children of the 1960s.[6] By the mid-1970s the modernist future was passé. Planners no longer proffered gleaming towers but "pictures of windmills and families holding hands," observed Banham; "what kind of future is that? Where's your white heat of technology? . . . Where's that homely old future we all grew up with?"[7] High-tech paradise perished in the wake of the Holocaust, Hiroshima, and postwar urban-development debacles. The rendezvous of twentieth-century dreamers with "that bright tomorrow turned out to be an appointment in Samarra."[8]

Yet that progressive future had itself been relatively novel. Prior to the

Enlightenment, Europeans had viewed past and future alike as ordained and predictable. Prognoses of times to come rested on the same religious chronology as annals of times past. History ran from the Creation to the End, whose coming was certain. The past was recounted and the future prophesied in definitive scriptural texts; "the Bible was not only a repository of past history, but a revealed pattern of the whole of history."[9] Human circumstances were thought constant over the entire sweep of mundane time. Since history was static, it could be exemplary; past, present, and future were wholly analogous. Repeated prophetic failures, such as predictions of the end of the world, merely postponed sacred prophecy. Each apocalyptic non-event increased the likelihood that the foretold End would come next time.[10]

This grand eschatological framework had little to do, however, with daily secular experience. Everyday affairs were beset by uncertainty. Save for regular cyclical diurnal and seasonal rounds, physical environment and social milieu were risky and insecure. Yet in principle, human and natural agencies, like divine ones, were presumed to be unvarying and predictable. The assumption of eternal sameness bolstered previsions of the future drawn from the past. Secular prognoses were based on exemplary historical evidence framed by a constant human nature; nothing really novel could arise. Whether the future was deduced from faith or by calculation, it was foreseeable because processes were unvarying. "He who wishes to foretell the future must look into the past," as Machiavelli put it, "for all the things on earth have at all times a similarity with those of the past."[11]

The constancy of the future was predicated not only on faith in the constancy of human nature and human agencies but on the general abhorrence of change common to most from Plato through the French Revolution. Although visions of desired stability might differ, "everybody equated happiness with absence of change," concludes the historian Peter Munz, "and considered change, even change for the better, to be intolerable."[12]

This traditional future, on the one hand comfortingly familiar, on the other depressingly foreclosed, gave way between the seventeenth and nineteenth centuries to the technological utopias described above. The primary impetus was the erosion of religious certitude by ideas and ideals of secular progress. While the new future was more confident and optimistic, it was less knowable and more mysterious than the traditional Christian morrow. Future forecasts were now transposed from the next world to this one; worldly experience, not sacred faith, now confirmed or denied human expectations. History was no longer simply divinely ordained but increasingly man-made and hence accessible to science and social engineering.

The pace of history also accelerated, notoriously at the end of the Napoleonic era, as observed by writers like Alfred de Musset and Gustave Flaubert.[13] After the French Revolution lives were disrupted with unique intensity. Europeans felt stranded between a past when change had been slow and life much the same from eon to eon, and a present that sundered each year from the last. Total rupture with the past, an unnerving fracture in time, was precisely what Jacobites had intended—a new order that would wholly expunge the old, from priests and patricians to weights and measures. Discontinuity was their deliberate legacy.[14] Social stability was one victim of the accelerated pace of change, intellectual security another. Unprecedented novelty eroded faith in the lessons of history. "What experience and history teach is this," as Hegel put it: "nations and governments have never learned anything from history."[15] As the present no longer predictably emerged from the past, neither could the future be foreseen in the present.

These new ways of viewing past and future, sacred and secular chronicle, did not come all at once nor entirely replace the old. As early as Luther, time's acceleration had seemed to bring forward the Last Judgment. Resisting such previsions as heretical, the papacy and the Holy Roman Empire clung to an annalistic conception of the past and a static view of the future, with sacred destiny repeatedly delayed. The last papal prophecy (1595) of the end of the world put it a long, safe time ahead—as far off as 1992.[16]

With human experience bereft of sacred constancy, the secular future became ever harder to ascertain. The next world's accelerated advent now became merged with the future of the existing world. Horizons of expectation shortened; people grew used to giving voice to what they wanted within their own lifetimes rather than delaying gratification to the hereafter. No prospective gain was any longer inconceivable; to many Rousseau's vision of the perfectibility of man seemed a reasonable, realistic historical agenda. But this wonderful future remained vague and formless.

Progressive change left the past ever less relevant as a guide to the future. As previous events lost their exemplary virtue, annals based on the cyclical regularities of stars and planets, rulers and dynasties, gave way to unique and open-ended narrative histories. What lay ahead was no longer preordained;

all that seemed certain was that "the future would be different from the past, and better, to boot."[17] But like past history, future events would be unique, erratic, irregular. Just as history now had to be explained anew by each generation, progressive change severed the future from anything anyone could anticipate.

The industrial revolution and European imperial expansion overseas seemed to confirm these previsions. But in the wake of technocratic hubris came growing fears about the future's social and cultural impacts, along with nostalgia for a past now felt to be irretrievable. The rupture of continuity following the momentous upheavals of the French Revolution, as noted above, inaugurated a pace of change felt psychologically and socially disastrous. Ominous harbingers generated new prophecies of catastrophe, now not divinely ordained but technologically caused. "The series of events comes swifter and swifter," judged Carlyle, "velocity increasing . . . as the square of time."[18] Anxiety over an unimaginable future culminated in Brooks Adams's prognosis of imminent societal dissolution.[19] For the first time in history progress seemed imminent, yet beyond that promise loomed a perhaps monstrous future.

Ambivalence toward that future and the supplanted past spurred the memorial occasions and commemorative icons that festooned European and American landscapes of the late nineteenth and twentieth centuries.[20] Their purpose was not just to remember the past but to commend it to future generations. Like a knotted handkerchief, plaques, flags, and tombstones are intended less to spur immediate recall than to fasten future recollection.[21] Without such reminders, heedless successors might take some catastrophic course. Only the persistence of the past could rein in a tearaway future. In country after country, heritage protection became a national creed embodied in legal codes to preserve the public patrimony against the vicissitudes of man, nature, and time.[22]

As environmental impact further eroded faith in technology, preservation sentiment expanded to embrace nature. Fears of annihilation unleashed by Hiroshima became widespread. As the unknown future grew ever more fearsome, mainstream scientists joined ecological gurus in dire warnings of a technological Armageddon. Multiplying pressures on the biosphere now presage incalculable damage. The public has learned to fear radiation and toxicity that mounts over time yet whose risk can be assessed only when precautions would be too late. Scientists are chided for failing to predict adverse effects with speed, precision, and certainty.[23]

Changes that may be irreversible are now felt to be most fearsome. Fears are not solely ecological; they bear on historic buildings and works of art, often "conserved" at the cost of quality and ambience. But impacts that risk the irrevocable loss of ecosystems are of paramount concern, for they threaten to extinguish human life, even all life.

The sheer magnitude of what is unknown makes today's future parlous. How much and what kinds of aerosol emission might irretrievably expand the ozone hole? How depleted can an ecosystem get before degrading beyond recovery? Slow bioaccumulation, the lengthy half-life of many radioactive disintegration products, the prolonged ecosystemic effects of species extinctions, the differential pace of various natural processes, the incommensurable acceleration of technological impact—all generate alarm about futures set in train by humans, yet whose outcomes humans cannot predict.[24]

The malign effects of progress today loom larger not simply because they seem more noxious and dangerous but because their attendant benefits have already been so long heralded. And as new conquests of nature come at ever greater expense, rising costs make it harder to resolve existing environmental problems or to respond effectively to new ones. Even when specific ill effects prove reversible, the environmental crisis is bound to persist. Mounting costs will make impacts harder to contain—especially when the Third World must cling for sheer survival to environmentally damaging technology.[25]

Hence progress now seems improbable on three counts. First, scientific enterprise costs too much to continue at the current pace. Inquiry into realms ever more remote from the macroscopic domain of everyday life, into the extremes of time and space, mass and temperature and speed, requires inputs of matériel and personnel exponentially greater than previous inquiries—inputs less and less justified by the benefits they yield. These economic and social impediments to further probing of the fundamentals of the firmament are bound to leave questions unresolved and, more distressing, questions unasked because unfathomable.

Second, what remains hidden would, were resources available to probe it, falsify much existing knowledge. Like the secure future of the religious faithful, the old unblinking confidence of the scientific community dwindles in the face of the stubbornly uncertain or unknowable. What we will probably never know now looms larger than the closely circumscribed realm of what we may be able to find out. Many if not most secrets of the universe will remain obscured.

Third, technology's unplanned and often unforeseen side-effects cause mounting disquietude. The risks of nuclear war, radiation, the greenhouse effect, the ozone layer, species depletion, and ecosystem loss haunt us because we are impotent to assess their magnitude, let alone to halt them.[26] Since Malthus, many have questioned whether science and technology enhance life. More and more now doubt it. Malign effects loom larger, not just because they seem more noxious but because their attendant benefits have already been discounted. Miracles publicized in advance can be letdowns when at length they come to pass. By comparison with achievements of the two centuries past and with science-fiction visions of the future, many of technology's new marvels seem humdrum, expectable, even trivial.

The social effects of manifold disillusionments—doubts that new scientific miracles will continue to rescue us from heedless greed, disbelief that technological progress will make people happy—are as disheartening as the foreshadowed physical failures. The collapse of inflated expectations and loss of faith in progress induce despondency, impotence, and *après-moi-le-déluge* escapism.[27]

The image of life as a career continuous from cradle to grave, which became a middle-class norm over the course of the nineteenth century, still governs Western narratives of personal identity. Future success is seen to demand self-fashioning; in the very act of planning ahead we aim to make something of ourselves.[28] Yet this careerist image now begins to look outdated. Recent decades have badly dented life-cycle expectations. Self-made young millionaires lack any vision of the rest of their lives. Social and political norms as well as scientific doubts preclude future expectations. People unsure what to expect are even uncertain what they ought to want to expect.[29]

Hence modern future orientations seem in some ways as schizophrenic as those of our medieval and Renaissance precursors. But our quite different future realms now engender utterly unlike reactions. When most people had little power to shape the circumstances governing their lives, proximate and private futures used to be fatalistically accepted. Because everyday conditions sharply curtailed the likelihood of change for the better, expectations about the future were postponed to the hereafter. But today many doubt any afterlife, much less a better one. As prospects for global improvement dim, hopes focus on mundane private futures. And these futures are increasingly short-term, bearing more on oneself and less on even immediate progeny. "The more gloomy we are," notes a British analyst, "the greater will be our concern for the present rather than for our grandchildren."[30] Longer-term futures are darkened by portents of decline, disaster, and chaos.

Diminished faith in progress is everywhere apparent. Science and technology are no longer guarantors of a better life. Parents no longer believe that their children will be better off—happier, richer, healthier—than themselves; to the contrary, they envision their offspring confronting ever graver problems. Such problems seem increasingly intractable; the more we extrapolate from the present, the less confident we are that we can prudently manage the future. The threats we face—notably environmental—seem complex and malign, the solutions to them partial and temporary at best. Many now consider it unsafe to tamper with nature at all even when they are well aware that they are powerless to stop doing so.

Three particular shibboleths fly in the face of earlier progressive visions. One is that nature is superior to culture and is at its best when least impacted and least controlled. The second is that tribal and indigenous peoples, in close rapport with life-enhancing nature, are ecologically less damaging than technologically advanced societies. The third is that nothing should be done that is irreversible, lest we foment some lethal terminus.[31] These retrograde views deny both the morality and the practicality of progress.

In sum, the future seems generally unappealing compared with the nostalgized past. It used to be said of planners that for them the past was when everything went wrong; in the future everything would be fine. Today these estimates are often reversed. I have termed the past a foreign country; the future is not a country at all but a chimera more often shunned than embraced.[32]

How did we come to this sorry state? What accounts for the now common surmise that progress, never as glorious as formerly touted, has been reversed and may now be precluded? Half a dozen environmental and cultural forebodings seem complicit in these gloomy prognoses.

1. *Fear of what seems unknowable.* The unknown is always a source of anxiety, but today's unknowns seem especially shadowed by uncertainty.

David Lowenthal

Modern analyses make the environmental future less predictable than ever; once presumed regular stochastic processes now seem at risk of chaotic derangement through megacosmic accident, episodic disequilibrium, or human impact. It is ecologically received wisdom that the long-term consequences of environmental changes already set in motion are more complex than we can ever foretell. Given the mutation pace of bacteria immune to antibiotics, no one now is so sanguine as the *Scientific American* editorialist of 1924 that the "natural outcome of the struggle between mankind and microbe has always favored mankind."[33] Faith that medical research augured "improvement without end" has given way to doubt, skepticism, public mistrust, and manifold fears.[34] Future security seems increasingly a mirage; we cannot be sure that we do what is best even for ourselves and our immediate heirs.

Historical change is similarly stripped of patterned certitude and regular progression. Past faith in the ultimate mastery of social goals gives way to anomie and drift. To be sure, science lengthens life spans, feeds ever more billions, globalizes markets and communications. But few readings of recent history chart progress toward liberty and freedom, education and welfare, even in the reduction of hunger, let alone of genocide or torture.

2. *The fetish of populism.* Progress was once aligned with increased democracy; those who chose their own rulers and made their own laws seemed apt to augment their own well-being. This supposition now seems dubious. Participatory democracy does mitigate the iniquities of autocracy, but it often fails to narrow, and may even widen, gulfs between rich and poor, powerful and impotent, as in most of the world today.

Moreover, democratic forms foment decision making that is lamentably short-term. Legislators elected every two or four years limit their vision to two or four years hence. Populist necessity rules out longer-term goals. A cult of instantaneity enthrones rule by opinion poll; tomorrow's polls overturn today's decisions. The sovereignty of the present moment makes stewardship a dead letter.

3. *Media hype.* The pace and pervasion of media dissemination reinforce the tyranny of immediacy. Because anything can be broadcast worldwide within seconds, today preoccupies us to the exclusion of yesterday and tomorrow. Emphasizing the present at the expense of hindsight and planning,

the media grabs attention by oversimplifying. Issues by their very nature nuanced, ambivalent, and culturally differentiated are purveyed as though they were simple, clear-cut, and universal.[35]

Everything is expressed in extremes. Only the paradisiacal and the catastrophic gain attention; and the latter, being more dynamic, dominates the headlines. The media world is one of general anxiety punctuated by episodic disaster. Oversimplification pleases a public that craves definite and succinct guidelines. "We want someone to say: If you ever clearcut on ash soils, the sky will fall in. If you ever cut trees on a slope steeper than 45 percent, the world will come to an end."[36] As with Chicken Little, the sky seems always about to fall.

4. *Mistrust of authority.* Vox populi proves a hollow promise. Empowered in principle by democratic institutions, the public is rendered impotent in practice by scientific doubts, by technological complexities, and by the anonymity of corporate dominion. Disillusionment and cynicism ensue. Opinion polls show steadily declining faith in government, science, and industry. Doubts mount not only of leaders' probity but of their ability to cope with issues even when honest. Those in charge of our fate, even those chosen by ourselves, seem ever more venal and ignorant.[37]

5. *Historical guilt.* We are increasingly bidden to shoulder blame for yesteryear's manifold iniquities. As the heirs of history's victims clamor for redress, the heirs of the victors offer token amends for ancestral misdeeds. African slavery and the Irish Famine attract formal apologies; mainstream spokesmen impotent to rectify present injustices readily regret those of the past.[38] Anachronistic guilt trips attest our superiority to benighted forebears while we deplore ancestral follies. In today's world of symbolic restitution, faith in progress comes across as repellent imperial hubris.

6. *Ignorance a palpable presence.* Increasingly we realize that education is not the panacea it was once supposed. Ignorance is not simply an absence of instruction but a substantive force, a mental feature instilled by teaching, often deliberately sought. We profess knowledge by dressing what is unknown or unclear in names, numbers, and moral absolutes. A world swamped by secular insecurity clings to mystiques of the sacred. Hence the rise of fundamentalisms that proffer transcendental absolutes at the expense of rational thinking.[39]

David Lowenthal

How can these retrograde tendencies be countered? Here are half a dozen suggestions:

1. *Welcoming change.* Both ecological and cultural history show that change is inevitable, whether wanted or not. Change is seldom cyclical, never reversible; each historical event brings the world to a state of being in some measure new. There is no way to revert to an earlier time, real or fancied. This being so, it is essential to acknowledge and make the best of it—not merely to mourn what has been lost or to deplore what replaces it but to welcome change as potentially promising. We must remember but cannot afford to repine for an irretrievable past.

2. *Conjoining conservation with creation.* A common feature of despair over progress is to sever a cherished past from a barren present, segregating protected heritage from unwanted novelty. This compartmentalization is futile. Nothing however protected endures forever. Every treasured material trace and relic continually decays and will finally vanish altogether, through either slow decomposition or sudden casualty—fire or flood, earthquake or war, iconoclastic revulsion or an excess of adoration. We ought not unduly regret such losses. For while attrition is inevitable, we go on revising and creating, adding new heritage to old. The stock of what is conserved and transmitted gets augmented as much as, if not more than, it is diminished. Rather than segregate what we conserve from what we create, we should recognize their kinship and rejoice in their intermixture.

Present-day concern with the past provides a cautionary warning for views of the future. Relics of memory and homage currently proliferate. But affection and protection sunder this glut of deliberate residue—antiques, mementoes, monuments, museums, historic sites, signposts—from the unregarded present. The preserved past might play a more vital role were it less memorial, more anticipatory.

Conservationists warn against jettisoning anything that some future might wish had been preserved; accordingly we ought dispose of nothing irreversibly. But this is a vain proscription; *every* present act forecloses myriad other prospects. The future cannot be foretold, but if it is to be reached at all, we are bound to relinquish some routes in favor of others.

Above all, we should aim to admire what we make, praising rather than disparaging our own cultural additions and environmental impacts. Taking pride in what we do is the first step toward truly praiseworthy actions. To regard our own productions as worth handing on, we need to feel that they are not merely inherited but our very own.

That values as well as relics are evanescent must also be borne in mind. The goods and goals our successors prefer will seldom be those we most cherish. We need to trust our heirs to make their own choices rather than to reaffirm our own. To this end, *our* handing down any particular stock of material goods matters less than *their* inheriting knowledge of craft skills, collective institutions in good working order, and faith in organizational resilience.[40]

3. *Stressing stewardship.* Stewardship is more preached than practiced. Economists calculate intergenerational equity but do not build it into decision making. We need to realize that stewardship is vital not only for our heirs but for ourselves. Without it human existence is sorely diminished, stripped of meaning and purpose. Faith in a past that we did not live in and in a future we will not experience is, as Durkheim showed a century ago, mandatory in all societies.[41] The protracted endurance, if not the presumed immortality, of the communities to which we belong shields us from the shallow loneliness of life's brevity, enlarging and extending it with collective memories and prospects.

Some deny the need to sacrifice immediate benefits for the sake of a probably more competent and fortunate future. A well-known economist scorns "the imposition of any burdens on people alive today . . . in order to add a few percentage points to the incomes of their far richer descendants towards the end of the next century [as] an anti-egalitarian form of inverted ancestor worship."[42] Solutions to shortages and risks that now seem intractable may indeed emerge. But this misses the spiritual point: only a heightened concern for future welfare lets us feel at home with our present selves. Awareness of the benefits of stewardship is not, however, automatically ingrained. It is a social value that has to be continually reinforced through community action and understanding.

4. *Lengthening future vistas.* The short-term immediacy that bedevils us today is not a consequence of democracy alone; it also reflects a lack of consensual pride in collective enterprise. Historical examples might energize our seemingly unheroic times: we need to put ourselves in other peoples'

places, other cultures' times. The founders of the American republic had a lively consciousness of their own historic importance. The framers of the Constitution were reminded by one 1776 pamphleteer that they were "painting for eternity" and so must be sure to get things right.[43] Mormons bent on the retroactive conversion of progenitors aim conscientiously to balance obligations to ancestors with legacies to descendants. Rituals of stewardship in many societies constructively bind up past with future.

To instill a like consciousness of enduring purpose we need to find or invent new enduring collective projects, echoing the prolonged completion of medieval cathedrals and the 999-year property leases of Victorian and Edwardian England.[44] The visionary Stewart Brand has built a gigantic mechanical clock that will record time for 10,000 years.[45] Some projects could be healing as well as heroic. We might address such long-term threats as the by-products of nuclear decay whose presence may be lethal for a million years. Indeed, such a crusade is essential if future generations are to inherit a viable planet. Active care for risks far in the future now engages environmental economists as never before.[46]

5. *Promote discourse.* We are beset by doubts that vitiate collective self-confidence save at the narrowest tribal level. While according each group equal credence and respect, we make little effort to learn what those unlike ourselves really think. Instead, a cult of essentialism valorizes apartheid, leaving each self-congratulatory constituency cocooned in its own solipsistic place.[47]

It is hard to engage in discourse with those from whom we differ when even exploring difference may be felt a slur. Confrontation with unlike minds might be encouraged through various media—reading aloud turn by turn, first-person historic interpretation, role-shifting theatrical performance. To achieve mutual understanding we must learn to refrain from proselytizing.

6. *Anticipating delight.* Progress in the best sense stems first from contemplating future pleasure, then from devising means of fructifying and enhancing it. Yi-Fu Tuan's writings are suffused with precisely this kind of foretold appreciation. Some thirty years ago he circulated a questionnaire asking those involved with environmental design what circumstances best promoted their anticipation of joy and pleasure.[48] In pursuing individual and often private delights we best equip ourselves to anticipate broader progress.

Our sense of the future might be made more tolerable by schooling ourselves not merely to forecast its possible perils but to dwell constructively on its at least equally likely pleasures.

Notes

1. Leo Marx and Bruce Mazlish, eds., *Progress: Fact or Illusion?* (Ann Arbor: U of Michigan P, 1997).

2. Patricia J. Williams, "The Slough of Despond," *Nation*, 9 March 1998, 10.

3. Bruce Mazlish, "Progress: A Historical and Critical Perspective," in Marx and Mazlish, *Progress*, 32. Texts from J. B. Bury's *The Idea of Progress* (1932) on are reviewed in Patrick McGreevy, *Imagining Niagara: The Meaning and Making of Niagara Falls* (Amherst: U of Massachusetts P, 1994), 103–7.

4. Reyner Banham, "Come in 2001 . . . ," *New Society*, 8 January 1976, 62–63. See also Michael L. Smith, "Recourse of Empire: Landscapes of Progress in Technological America," in *Does Technology Drive History? The Dilemma of Technological Determinism*, ed. Merritt Roe Smith and Leo Marx (Cambridge: MIT Press, 1994), 37–52.

5. Olaf Stapledon, *Last and First Men: A Story of the Near and Far Future* (1930; reprint, Harmondsworth: Penguin, 1987), 15.

6. Mary McCarthy's *The Oasis* (New York: Random House, 1949) was a paradigmatic preview of 1960s eco-communes.

7. Banham, "Come in 2001 . . ."

8. Bruce McCall, "A Glimpse Back from 2050," *New Yorker*, 20–27 October 1997, 270.

9. Yosef Hayim Yerushalmi, *Zakhor: Jewish History and Jewish Memory* (Seattle: U of Washington P, 1982), 21.

10. Frances Carey, ed., *The Apocalypse and the Shape of Things to Come* (London: British Museum Press, 1999).

11. Niccolò Machiavelli, *The Discourses*, trans. Leslie J. Walker, 2 vols. (London: Routledge & Kegan Paul, 1975), 1:575; originally published as *Discorsi* (1532).

12. Peter Munz, *Our Knowledge of the Growth of Knowledge: Popper or Wittgenstein?* (London: Routledge & Kegan Paul, 1985), 314.

13. Richard Terdiman, "Deconstructing Memory: On Representing the Past and Theorizing Culture in France since the Revolution," *Diacritics* 15 (1985): 13–36; idem, "The Mnemonics of Musset's *Confession*," *Representations* 26 (spring 1989): 26–48.

14. Roy Pascal, *Design and Truth in Autobiography* (London: Routledge & Kegan Paul, 1960), 57; François Furet, "L'Ancien Régime et la Révolution," in *Les lieux de*

mémoire: III. Les France: 1. Conflits et partages, ed. Pierre Nora (Paris: Gallimard, 1992), 113–19.

15. G. W. F. Hegel, "The Varieties of Historical Writing" [1822], in his *Lectures on the Philosophy of World History*, trans. H. B. Nisbet (Cambridge: Cambridge UP, 1975), 21; originally published as *Vorlesungen über die Philosophie der Weltgeschichte* (1837).

16. Reinhart Koselleck, *Futures Past: On the Semantics of Historical Time*, trans. Keith Tribe (Cambridge: MIT Press, 1985), 9.

17. Ibid., 6–18, 32–38, 280.

18. Thomas Carlyle, "Shooting Niagara: and After?" [1867], in his *Critical and Miscellaneous Essays*, 3 vols. (London, 1887–88), 3:590.

19. Brooks Adams, *The Law of Civilization and Decay* [1896], 2nd ed. (New York: Vintage, 1955), 292–95, 307–8.

20. Eric Hobsbawm, "Mass-producing Traditions: Europe, 1870–1914," in *The Invention of Tradition*, ed. Eric Hobsbawm and Terence Ranger (Cambridge: Cambridge UP, 1983), 263–307. See also John Bodnar, *Remaking America: Public Memory, Commemoration, and Patriotism in the Twentieth Century* (Princeton: Princeton UP, 1992); John R. Gillis, ed., *Commemorations: The Politics of National Identity* (Princeton: Princeton UP, 1994); and "Studies in Representations of the Past," special issue of *History & Memory* 5, no. 2 (1993).

21. Alan Radley, "Artefacts, Memory, and a Sense of the Past," in *Collective Remembering*, ed. David Middleton and Derek Edwards (London: Sage, 1990), 46–59.

22. I trace this development in *The Heritage Crusade and the Spoils of History* (London: Viking, 1996), 3–11.

23. Samuel P. Hays, *Beauty, Health, and Permanence: Environmental Politics in the United States, 1955–1985* (Cambridge: Cambridge UP, 1987), 182–84; Kai T. Erikson, *A New Species of Trouble: Explorations in Disaster, Trauma, and Community* (New York: Norton, 1994); Maurie J. Cohen, "Risk Society and Ecological Modernisation," *Futures* 29 (1997): 105–19.

24. Alan Randall, "Human Preferences, Economics, and the Preservation of Species," in *The Preservation of Species: The Value of Biological Diversity*, ed. Bryan G. Norton (Princeton: Princeton UP, 1986), 86–87; Leo Marx, "The Idea of Technology's Postmodern Pessimism," in Smith and Marx, *Does Technology Drive History?* 237–57.

25. World Commission on Environment and Development, *Our Common Future* (Oxford: Oxford UP, 1987); Nicholas Rescher, *Unpopular Essays on Technological Progress* (Pittsburgh: U of Pittsburgh P, 1980); Ramachandra Guha and Juan Martinez-Alier, *Varieties of Environmentalism: Essays North and South* (London: Earthscan, 1997).

26. Rescher, *Unpopular Essays on Technological Progress*, 282.

27. Ibid., 19, 24–28.

28. Alan Dundes, "Thinking Ahead: A Folkloristic Reflection on Future Orientation in the American Worldview," *Anthropological Quarterly* 42 (1969): 53–72; Jonas Frykman and Orvar Löfgren, *Culture Builders: A Historical Anthropology of Middle-Class Life* (New Brunswick: Rutgers UP, 1987), 29–30.

29. Gunnhild O. Hagestad, "Dimensions of Time and Family," *American Behavioral Psychologist* 29 (1986): 688.

30. Raymond Plant, "What Has Posterity Done for Mrs Thatcher?" *Times* (London), 10 July 1989, 14. See also Kenneth Blaxter, *People, Food, and Resources* (Cambridge: Cambridge UP, 1986), 97.

31. I discuss these shibboleths in "Environmental History: From the Conquest to the Rescue of Nature," in *Cultural Encounters with the Environment: Enduring and Evolving Geographic Themes*, ed. Alexander B. Murphy and Douglas L. Johnson (New York: Rowman & Littlefield, 2000), 177–200. See also George W. Stocking Jr., "Rousseau Redux, or Historical Reflections on the Ambivalence of Anthropology to the Idea of Progress," in Marx and Mazlish, *Progress*, 65–81.

32. David Lowenthal, *The Past Is a Foreign Country* (Cambridge: Cambridge UP, 1985); G. Rattray Taylor, *How to Avoid the Future* (London: Secker & Warburg, 1975).

33. Quoted in Nancy Tomes, *The Gospel of Germs: Men, Women, and the Microbe in American Life* (Cambridge: Harvard UP, 1998).

34. Leon Eisenberg, "Medicine and the Idea of Progress," in Marx and Mazlish, *Progress*, 45–64.

35. Gregg Easterbrook, *A Moment on Earth: The Coming Age of Environmental Optimism* (New York: Penguin, 1996), 464–65.

36. Nancy Langston, *Forest Dreams, Forest Nightmares: The Paradox of Old Growth in the Inland West* (Seattle: U of Washington P, 1995), 285. See also M. Jimmie Killingsworth and Jacqueline S. Palmer, *Ecospeak* (Carbondale: Southern Illinois UP, 1992).

37. Joseph W. Nye Jr., Philip D. Zelikow, and David C. King, eds., *Why People Don't Trust Government* (Cambridge: Harvard UP, 1997); Brian Wynne, "May the Sheep Safely Graze? A Reflexive View of the Expert-Lay Knowledge Divide," in *Risk, Environment, and Modernity: Towards a New Ecology*, ed. Scott Lash et al. (London: Sage, 1996), 44–83; Rolf Lidskog, "Scientific Evidence or Lay People's Experience? On Risk and Trust with Regard to Modern Environmental Threats," in *Risk in the Modern Age*, ed. Maurie J. Cohen (London: Macmillan, 2000), 196–204; Michael R. Edelstein, *Contaminated Communities: The Social and Psychological Impacts of Residual Toxic Exposure* (Boulder: Westview, 1988).

38. Roy L. Brooks, ed., *When Sorry Isn't Enough: The Controversy over Apologies and Reparations for Human Injustice* (New York: New York UP, 1999).

39. Paul Boyer, *When Time Shall Be No More: Prophecy Belief in Modern American Culture* (Cambridge: Harvard UP, 1992); Eugen Weber, *Apocalypses: Prophesies, Cults, and Millennial Beliefs through the Ages* (Cambridge: Cambridge UP, 1999).

40. Mary Douglas and Aaron Wildavsky, *Risk and Culture: An Essay on the Selec-*

tion of Technological and Environmental Dangers (Berkeley: U of California P, 1982), 197–98; Allen Tough, "What Future Generations Need from Us," *Futures* 25 (1993): 1041–50.

41. Emile Durkheim, *The Elementary Forms of Religious Life,* trans. Karen E. Fields (New York: Free Press, 1995), 213–14, 351–52, 372, 379; originally published as *Les Formes élémentaires de la vie religieuse* (Paris: F. Alcan, 1912).

42. Wilfred Beckerman, "Warming to Global Change," *Times* (London), 11 December 1997; see also idem, *Small Is Stupid* (London: Duckworth, 1995).

43. Quoted in Cynthia S. Jordan, " 'Old Words' in New Circumstances: Language and Readership in Post-Revolutionary America," *American Quarterly* 40 (1988): 501.

44. David Remnick, "Future Perfect," *New Yorker,* 20–27 October 1997, 210–18.

45. Stewart Brand, *The Clock of the Long Now: Time and Responsibility* (London: Weidenfeld & Nicolson, 1999).

46. Andrew Blowers, "Nuclear Waste and Landscapes of Risk," *Landscape Research* 24 (1999): 241–61; Paul R. Portney and John P. Weyant, *Discounting and Intergenerational Equity* (Washington, D.C.: Resources for the Future, 1999).

47. Walter A. McDougall, "Whose History? Whose Standards?" in *Reconstructing History,* ed. Elizabeth Fox-Genovese and Elisabeth Lasch-Quinn (New York: Routledge, 1999), 282–98.

48. Yi-Fu Tuan to the author (and others), 1 August 1967.

Chapter 4

Progress and Anxiety

Yi-Fu Tuan

In the early 1960s the distinguished British historian J. H. Plumb planned a series of volumes of historical studies with the overarching thesis that "the condition of man now is superior to what it was." A dozen years later, in 1977, Martin Mayer noted that Plumb's "innocuous and obvious factual statement would draw looks of incredulity and perhaps outrage, especially from the young. There is abroad in the world—not only in America—a need to believe that everything is getting worse. . . . English youngsters who thrill to a Sherlock Holmes scene set in an impenetrable London fog will talk earnestly about the worsening air pollution of a London that hasn't seen such a fog in twenty years."[1]

Little has changed since Mayer's remark. Today, among the educated young and among liberals generally there remains a deep-seated, almost visceral suspicion of the idea of progress: every improvement in environmental quality, housing, hygienic practice and life expectancy, literacy, or population control is either controverted, or taken to be severely restricted in terms of its area of application, or seen as having dire long-term consequences. Strange to say, premonitions of disaster are most evident, not in the poorer countries, but in the wealthiest and technologically most advanced countries of the world, foremost among them the United States. Why is this? There must be local reasons, reasons applicable to the United States above all. I shall discuss one such reason, but to do it justice and put it in context, I need to paint in broad strokes first. I shall start with anxiety caused by change as such—any change, even change for the better.

The Anxiety of Growing Up

One change is acknowledged everywhere to be for the better—growing up. Parents eagerly register the progress of their child. An infant is immobile. By its sixth month it can crawl, and it does so with growing confidence until one day it struggles to stand up and walk. Why it should want to do so is not immediately obvious, for it is safer, speedier, and more efficient to continue to crawl. Standing up—defying gravity—entails a new risk. Walking on two small feet is unstable, and the child periodically falls. Yet the child will stand up, with pride but also with a tinge of anxiety. At the mental or psychological level a newborn infant is without fear. At a later stage it will show fear, including fear of the dark. What is so bad about the dark in a young child's sheltered life? Nothing real, but the child's maturing imagination learns to populate it with monsters. Imagination is a wonderful gift that humans possess to a unique degree.[2] Why is it not more common in the animal kingdom? Could it be because the ability to imagine—to see what is not, or not yet, there—is an ambivalent gift? Imagination is empowerment: it opens up all sorts of possibilities for action that ensures a safer and better world. Unfortunately, the world it opens up can also be bewildering and frightening. A child, as his or her body and mind mature, grows in confidence yet also, contradictorily, in anxiety and doubt.

The child is not left adrift. Society comes to his aid. Others before him have confronted similar anxieties and doubts and have devised answers to the child's questions—have at least posed the same questions, which reassures him that he is not alone. Answers are given in the form of stories, myths and rituals, customs and codes of behavior. These make up a mental edifice that provides a haven against anxieties and doubts, just as the physical edifice of houses and fenced-in fields provides a haven against bad weather, wild animals, and human enemies. Culture thus serves a double purpose: it demarcates worlds that offer protection from external forces and domesticates people's unruly imagination by confining its scope to previously explored spaces and channels. Put this way, culture is made to sound conservative. Yet, viewed from a different angle, culture is dynamic, the product of a probing imagina-

---

tion exercised by people over a long period of time for the purpose of extending their power. As such it will have a history, the general direction of which is from exploratory beginnings to refined achievements. A term for such directional growth is *progress*.

Every human group has known progress: however simple its culture, it could only have been achieved by some cumulative process. Yet few human groups known to history have recognized and welcomed the idea of cumulative change, perhaps because by opening up the possibility for change it reminds human beings of the conditionality and uncertainty of everything they have. Nature is wayward enough. To see that culture too is unstable and subject to radical transformation would be unsettling in the extreme. But what about change into something better? Well, historically, very few people could see, or had reason to see, the future in a confident light. If a better world was in the offing, the gods or culture heroes would have to intervene, as they had done at the beginning, in sacred mythical time. Surprising as it might seem to us, people in the past believed that it was not they but the gods and ancestral heroes who produced culture. The task of human beings was to maintain and reproduce it; and they knew only too well that in this their success was mixed at best, for even in a well-functioning community stresses and failures were common facts of life. People saw, then, a need for periodic renewal, and that meant a return—a journey backward in time—to wholesome, mythological beginnings.[3]

Progress—At Last and Alas!

Although rare, the idea of progress has made its way, as a subsidiary theme, into certain cultures that have embraced other views, for example, the view of decline from a golden age to ages of baser metal or a view of cyclical oscillation. The ancient Chinese and Greek cultures are cases in point. How is the subsidiary idea of progress expressed in them? I would like to cite certain reservations and anxieties discernible in Chinese and Greek ideas of progress to argue that such reservations and anxieties are universal: they exist wherever the idea of progress exists. Even the modern West's profession of progress, by far the boldest and most confident in the world, is not exempt from nervous glances over the shoulder.

In China, the idea of progress, other than as an individual adept's rise to immortality, is absent from Taoism. This is understandable, for Taoism despises civilization. Confucianism, by contrast, favors civilization. True, it believes in a golden age in the past, which would seem to rule out progress, but its golden age, unlike that of Taoism, was attained not in nature but in the civilized states of antiquity—the Hsia, the Shang, and the early Chou. Confucius himself suggested that the last of the three—the Chou—represented a peak because it could benefit from the experience and accomplishments of its predecessors.[4] Progress of a political and ethical kind was implied. What about material progress? Although thinkers such as Mencius and Ssu-ma Ch'ien were able to praise the economic wealth generated by vigorous merchant entrepreneurs of their time, they did not conceive of civilization in material terms. Not even Mo-tzu, a thinker who came from the artisan class. Mo-tzu, who, unlike Confucius, could boldly conceive of a future state of universal love, was unable to imagine a future state of material abundance and architectural grandeur. He spoke admiringly of the sage kings of the past who made do with houses just high enough to avoid the damp and sturdy enough to withstand the onslaughts of nature. Houses, in other words, were to be erected from necessity, and not for show and prestige. Civilization meant civility, good manners, and the arts rather than the construction of a splendid material world.[5]

Anxiety over Material Progress

Large and sophisticated societies—those, for example, in historical East and South Asia, the Mediterranean region, and Europe—have radically transformed nature, creating whole countrysides of permanent fields, farmsteads, villages, and towns. Until modern times, however, such societies seldom boasted of their achievement.[6] Why is this? One reason is that fields and farms were historically identified with country folk and the peasantry, a class branded with the stigma of powerlessness. Who, then, did have power over nature? Who was able to force it to yield food dependably? The answer is, until modern times no one, though there was no lack of monarchs and potentates who claimed that they were generative by nature, that they had the *natural* powers to induce fertility.[7] Their claim was one thing, but what was the reality—the day-to-day experiences of people who lived on the land, managed it, and worked on it? Well, it was harsh. Until the eighteenth century, even in

---

the best-endowed parts of western Europe, farmers could see no clear causal relationship between their effort and success. A good harvest, or just an adequate harvest, was a matter of fate or God's will. As John Calvin put it, "Nor do we believe, according as a man will be vigilant and skillful, according as they have done their duty well, that they can make their land fertile; it is the benediction of God which governs all things."[8]

Historically, the search for power—for some sort of control over the way-ward terrestrial forces that intimately affected human livelihood—was to look skyward. If only the regularity and majesty of the cosmos could be brought down to earth. If only humans had the imagination and boldness to tap the power that lay in heaven and was invested in its gods, and acquire some of it for themselves. This skyward reach aroused anxiety. It implied (to use the Greek term) hubris, or excessive pride, on the part of mere mortals.[9] Such a reach would offend both the spirits of the earth and the Sky God and would perhaps invite punishment from them in the form of catastrophic collapse. To ensure that nothing dire happened, human builders of the cosmic city resorted to making offerings—including, in early days, human sacrifice—to the deities. Consider the capital of Shang China (ca. 1500 B.C.). It was in-tended to be a city for the living, but it might as well be called a necropolis for all the dead bodies—human sacrifices—that archaeologists have unearthed. Sacrificial victims were buried beneath every important edifice; indeed, every pillar of an important edifice might have a human body underneath.[10] The ancient Hebrews, to take a people from a totally different time and culture, also had their doubts about the propriety of overweening ambition. In Gene-sis we read that when Noah's descendants attempted to build a city in the land of Shinar and thus bring heaven to earth (or earth to heaven) by their own strength, the jealous God intervened. He not only dispersed the builders but made them speak in mutually incomprehensible tongues so that they could never congregate in sufficient numbers to form a great and proud society that owed little or nothing to God.

Every attempt to rise above our earthbound condition is a risk, invit-ing a fall. The earliest dwellings hugged the ground, were in fact semi-subterranean. As construction skills improved and confidence increased, houses were built above ground and of rectangular rather than rounded shape. The more important ones stood on platforms. In general, the more important the building, the taller it was. Sacrifices were made to assuage anxiety over the presumption of building grandiosely. Note how even today a groundbreaking ceremony precedes the construction of a major building and a topping ceremony marks its completion. Groundbreaking! How dare one, even today, tear the flesh of mother earth and burden it with a human tower without some propitiatory gesture.

The history of architecture in the West is the history of creating an artificial world that has progressively distanced itself from nature. Every step along the way has aroused both pride and unease. Very late in the game, when gaslight enabled Europeans to "conquer night" for the first time, a Cologne newspaper in 1816 did not hesitate to bring up the old argument that it transgressed "the divine plan of the world."[11] In the second half of the twentieth century, despite attempts to reintroduce tokens of nature back into buildings and building complexes, the overall trend toward artifice continued, indeed accelerated. The story is a thoroughly familiar one. For example, before the energy crunch of the 1970s Americans often found it necessary to put on a sweater in summer and strip down to a lightweight shirt in winter, so radically had the technology of cooling and heating overruled the seasonal impact of nature. In skyscrapers, as we well know, many offices have no windows and depend wholly on artificial lighting. The shopping malls that dot the landscape are covered Edens in which customers can buy, eat, go to the movies, and (in some) get married, in total disregard of nature's moods, both benign and horrid, outside. Life in such places, for all their sparkle and smooth efficiency, and even on account of them, can seem (after a while) unreal, a floating bubble or dream from which one will surely be rudely wakened.

A striking example of this persistent unease over technological prowess is captured in America's enduring fascination with the story of the *Titanic*.[12] A small library of books and at least four movies have been devoted to the fateful ship. In a review of the most recent remake of the movie *Titanic*, which packed theaters the world over in 1998, the critic Stanley Kauffmann wrote: "The luxurious *Titanic* was called unsinkable, the safest ship ever built; and it went down on its maiden voyage in April 1912, four days after it had sailed from Southampton to New York. Within a few hours of hitting an iceberg, dabs of caviar on special dinner ware had given way to life jackets in the freezing Atlantic. The contrast is hypnotic. . . . The whole story is a relentless reminder that death is always there pawing at the portholes and that it takes very little—one miscalculation in this case—to render the most resplendent order into primal chaos, then into nullity."[13]

Is the whole earth, now linked in all its parts by instant electronic communication, a proud ship sailing through the solar system with all its lights blazing? The widespread dread of catastrophic collapse in computer capability as the millennium approached month by month, week by week, and finally day by day in 1999 is the latest dramatic example of the human ambiguity toward technical achievement. As Oscar Wilde might have said, "No genuine progress should go unpunished."

Anxiety over Personal Fate

I noted earlier that both the ancient Chinese and the ancient Greeks entertained the idea of progress. I then used the Chinese version, with its stress on manners rather than material goods, to argue that societies generally, and not just Chinese society, have a tendency to harbor, in the depth of their collective psyche, an ambivalence toward life that has moved too far above its earthy roots. I now turn to the Greeks for another source of dissatisfaction with progress. In a famous passage in Sophocles' play *Antigone* the choral ode declaims how the arts and the sciences have brought human beings, step by step, from helplessness to a mastery of nature and to their crowning achievement, which is the sociopolitical state. Technical knowledge, the ode seems to suggest, makes it possible for human beings to make themselves more immune to luck—to just what happens to them as opposed to what they can make happen. Unfortunately, the immunity has clear limits. Contingency remains a daily threat, especially to individuals, and there is no victory over death.[14]

Is the situation any different today? Objectively, yes. We continue to gain control, which means that we are less and less subject to luck—to the whims of nature. It is useful to remind ourselves how insecure life was in an earlier time, even among the well-to-do. Consider Ralph Josselin, an English clergyman-farmer of the seventeenth century. His annual income came from three sources—his living as vicar, his farm, and his job as schoolteacher. One might expect him to feel reasonably secure. Yet throughout a diary that he kept for more than forty years one finds the refrain of accident, pain, and death. His diary gives the impression that he and his family were playthings of fate. Unseasonal rain or drought posed a recurrent threat to his crops and could quickly endanger his livelihood. A fire might kill his children and make him a pauper overnight, for there was no fire insurance in his time. In the absence of modern cures even a small mishap such as pricking one's thumb with a thistle could lead to gangrene and painful death.[15] How protected we are by comparison, not only by technical skills and the products of technology, from central heating to heart transplants, but also by such social instruments as insurance, both governmental and private.

A family farm in twenty-first-century Wisconsin is undoubtedly far more comfortable and secure than one in Josselin's time. Progress has been made. Yet even when recognized, it is seldom given its due. I have already offered one reason, namely, fear of rising above our human station and thereby offending the gods. Let me add three others that have a more secular flavor. One is the ease with which we accept our improved state and forget the hardships of an earlier time. Another is a consequence of progress itself: being able to see farther can be a source of anxiety. True, we gain greater control, but seeing farther means that we are simultaneously aware of that which we do not yet control. And thirdly, as individuals we remain subject to luck. Accidents will happen. I can still be killed by lightning or a drunken driver. And so in early-twenty-first-century America it is not altogether surprising that the gambler's dice swing under the mirrors of souped-up automobiles, although it is surprising, indeed shocking, that in the White House itself, that ultimate center of power and control, the course of events could seem so uncertain that a recent First Lady (Nancy Reagan) felt the need to consult an astrologer.

Strange to say, safety and predictability can make us feel anxious. Uncertainty is so much a part of our collective memory—reinforced, perhaps, by grandfather's hair-raising tales—that it can seem unnatural to have our projects (other than routine ones) come out right. A trip to Florida, or even further afield, to Tibet, without mishap may even be vaguely disappointing, as though it were not quite real. Unknown to ourselves, we may have hoped for a flat tire in the Everglades, or a canceled hotel reservation in Lhasa, to add an element of surprise to our adventure. After all, what is adventure without it? How can a good story be made of our trip without an enlivening mishap? Something else may be at work in this strange, unadmitted and inadmissible desire for a minor accident or two in our life course and plans, namely, the superstitious belief that they may appease fate and so postpone the major accident that will one day permanently maim or kill us.

Moral Progress and Its (Unwelcome?) Challenges

Progress is ultimately fatuous or empty unless it contributes to moral and intellectual awareness. Let's explore these two kinds of awareness, which in secular terms may be said to lie at the heart of human dignity.[16] Let's start with the moral. Has there been progress? I say yes, if only because major human attainments necessarily have a past, sometimes a lengthy one. Individuals gifted with moral sensitivity may be born any time and in any place, but their gifts are likely to remain merely potential without the support of a moral ethos, backed by a more or less articulated philosophy, that is already in the air. Such an ethos cannot be just an overnight inspiration. Moreover, evidence from the past shows that moral ideals and practices emerge and develop not all on their own but in tandem with material progress and growth in institutional scope and efficiency. They require these material and institutional supports to survive and spread.

Now, morals are customs, which are infinitely varied. How can one even begin to say which morals represent an initial practice and which represent a later expansion and refinement, which are crude superstitions and which are genuine achievements? One cannot, that is, unless one moves, as I shall, to a higher level of abstraction. Rather than addressing particular customs, I draw attention to the universal custom of reciprocity and show how, out of this universal custom, another, more demanding and elevating one emerged. This new morality, having emerged at a certain time and place, was not, for a time, universal. Yet, once it was articulated, its appeal to both reason and desire was sufficient to ensure its acceptance, as a goal if not in immediate practice, by people who had reached a certain stage of material and institutional development almost everywhere.

Reciprocity means that when one person helps another, that other is obliged to return the assistance, sooner rather than later. Reciprocity is thus a form of cooperation, and it goes without saying that human beings—and, for that matter, animals—must cooperate to survive. First and foremost, reciprocity is practiced among people who are geographically close. This means kinfolk and neighbors who, though unrelated by lineage, are given honorary membership in the family for purposes of mutual assistance. Moral obligation stops with the group, for only its members live close enough to give help on a daily basis and in emergencies. Given limited resources, most communities—in the past, and even now in backward areas—cannot afford to see the neediness of outsiders, much less help them. Another force at work toward the same end is this: when outsiders *are* noticed, it pays to see them in a somewhat hostile light, for such perceived hostility not only exonerates one from any obligation to help but also serves to strengthen the group's own internal cooperativeness and integration.

Throughout much of human history group loyalty was paramount: interest in the outsider was pretty much confined to relationships of trade that benefited both parties. The big step forward in moral thought—a step that could seem at first unnatural and inhuman—was the ideal of common humanity, with implied obligations far transcending one's own group, which first appeared in the world between 600 and 300 B.C. Buddhism, Stoicism, and, in China, the teachings of Mo-tzu all took this universalism to be human nature, though one would not think so to look at actual practice.[17] At a later time, Christianity famously redefined *neighbor* to mean anyone one happens to encounter, at home or abroad. Such a neighbor may need help, and it is to be given even when there is no chance of any return, now or later. Reciprocity—a sort of "you scratch my back and I'll scratch yours" morality at its crudest—turns into more open and generous circular giving: A gives to B, B to C, C to D, and D—who knows?—may one day give to A.[18] The circle of giving expands as society itself gains efficiency and confidence, so that eventually much of giving is linear. An individual or group passes on its surplus of energy and resources to contemporaries wherever they happen to be and to future generations. The reward of such giving is the reward of having responded to the call of humanity.

In the early hours of August 17, 1999, a major earthquake and its aftershocks killed tens of thousands in Turkey. It was the most disastrous earthquake in Europe in a century. Almost immediately nations—ultimately some sixty—came to Turkey's rescue. Aid arrived from governments, private charities, and individuals. So much aid poured in so fast that for a time it created its own disabling chaos. Television crews reported daily the rescuers' heroic attempts to dig out the buried, and when a survivor emerged out of the rubble, especially when it was a child, the rejoicing seemed a palpable sensation shared throughout the world. What a shock to modern sensibility it would be if the world sat on its hands! We take the widespread response for granted. But consider a natural disaster in Han dynasty China—say, in the year A.D. 100. Would the Roman emperor and the ruler of the Parthian empire have jumped to the rescue? Would their citizens and subjects have sent gold and blankets? The questions are, of course, rhetorical, raised to dramatize an easily forgotten point. Universal brotherhood may have been preached centuries before A.D. 100, but it was heard and practiced by only a few—outstandingly, those in some sort of religious order. Most people would have dismissed the ideal as hare-brained. Moreover, and just as important, there was no practical means to give help to victims so far away even if the wish to do so were there. Charity to strangers is more closely linked to technological advance than we care to admit.

Now, why would this expanded sense of moral obligation produce anxiety? Why can it make people feel vulnerable, so much so that they yearn for a return to narrow reciprocity, to a charity confined to kinfolk and neighbor? I offer a couple of reasons. One is the feeling, common among people in our hi-tech society, that no one is around to hold *my* hand or offer *me* chicken soup when I need comfort. I am alone, on my own. True, if I were struck down by a car in an enlightened and well-run city, I could count on immediate and efficient help. For a time I would be treated by total strangers as their brother: traffic cops would stop cars to let the ambulance through; at the hospital emergency room, doctors and nurses would probe and touch me solicitously. But once they saw that I was not badly hurt, and once they had signed me out, I would be, well, on my own. The nurse who held my wrist to feel my pulse only five minutes earlier would now look through me as though I didn't exist.

The second reason is the burden of choice—a dis-ease of affluence. Obligations to my family remain: bills from school and the orthodontist still have to be paid. On the other hand, television so effectively brings tragedy, with all its sights and sounds, right into one's home that the compulsion to write a check—but for how much?—to the earthquake victims of Turkey is irresistible. Irresistible, yes, but the gesture is always accompanied by questions of priority, of just where one's obligation stops. What about a neighbor's eloquent plea for animal rights?[19] When he comes to solicit money for his cause, can I say no, that the upgrading of my VCR to DVD must come first? The modern man and woman are forced to think morally, whether they want to or not.

Intellectual Progress and Its (Unwelcome?) Challenges

I now turn to intellectual progress—to the broadening and deepening of the human mind. Surely, this is a good thing? Yet it has its detractors. It has been accused, for example, of making the world seem grayer and emotionally anemic. Intellectual progress can have this effect on most people. But this is because most people lack the stamina and the talent to go all the way, beyond the thinning and the grayness, to reality's center or core, which may once again be an emotionally charged beauty.

Let me use our knowledge of the heavens to illustrate this progression. For a long time the Ptolemaic model of concentric spheres with the earth at the center reigned. It was eventually displaced by Copernicus's model, which put the sun at the center. From a purely scientific viewpoint this was a vast improvement—progress in simplification and elegance as well as in accuracy. But from an artistic and literary viewpoint the displacement of the Ptolemaic model meant the end not only of a scientific or empirical system but also of a magnificent, imaginative edifice whose foundations had been laid in antiquity and then built on throughout the Christian millennium. Its elements included grandeur—the cosmos was conceived as immense, though also bounded, like a Gothic cathedral, and therefore reassuring. Its transparent spheres were made visible by the stars embedded in them, and the stars were shining "intelligences," not dull matter, as we now see them. The rotation of the spheres created heavenly music. Earth, located at the bottom of the celestial hierarchy, might well be drab, but its human dwellers were not confined to it. They could look up and, looking up, be able to see successive layers of heaven leading to the realm of the Unmoved Mover, or God, a realm suffused in pure intellectual light and love.[20]

Given the beauty of this cosmos, its grip on the European imagination right into the eighteenth century is understandable. People who supported a courtly ideal of the state and people of a literary and artistic bent especially regretted its fading. A most poignant expression of loss was made by someone as noted for his mathematical gifts as for his literary sensibility—Blaise

Pascal—who famously wrote: "The eternal silence of these infinite spaces terrifies me."[21] For the next two centuries Pascal's older contemporary Isaac Newton provided the dominant paradigm. His cosmos, absent the rotating globe and music, was much grayer. On the other hand, it retained one beauty: lawfulness. People who could not follow the mathematics of Newtonian science could nevertheless rest assured, on Newton's authority, that all was stately and right in heaven. Now, at the beginning of the twenty-first century, we don't have even that. The universe, according to the current view, is violent. Besides the initial big bang, sudden explosions of high-energy radiation (gamma-ray bursts) occur almost daily at random positions in the sky. Our own Milky Way, far from being an uneventful progression from a primordial gas cloud to an orderly swirl of stars, had a violent youth, during which it tangled with and gobbled up several smaller galaxies in its neighborhood. As for predictability, Newton had it right for the solar system—but not quite. One-hundred-million-year simulations of the orbits of the planets reveal that all nine are chaotic. "The slightest change in a planet's initial conditions of position or velocity makes its motions millions of years hence entirely unpredictable."[22]

Steven Weinberg, a Nobel laureate in physics, is among those who can see that the "orbits of the planets are the results of a sequence of historical accidents" and so are "not beautiful." Nevertheless, for him, beauty exists, and it is to be found when he and other physicists study "truly fundamental problems."[23] But this is where idealistic thinkers run into trouble. They have the noble desire that great beauty be accessible to everyone. Well, it just isn't. There was a time when any educated European could enjoy the visual *and* auditory harmony of the universe. As science progressed, it presented a universe that was less and less sensorially and intuitively appealing. Only to the mathematically talented and highly trained—people capable of studying "truly fundamental problems"—will the universe remain mysteriously beautiful.

The broadening income gap between rich and poor in the last three decades is rightly a cause for concern. But what about the broadening intellectual gap? It is unconscionable that most of us can hold on to beauty only when it is based on delusion (for example, the idea that the universe is made up of musical spheres or that dancing in a circle produces rain), whereas a tiny few can hold on to both beauty and truth. The problem comes to a head in the

United States, for it is a country that has made remarkable scientific progress in its short history, that believed and still believes in progress, but that also embraces (especially among its idealistic young) a sort of radical egalitarianism.

Progress and Anxiety in the United States

A great American myth of progress, which also happens to be reality if we view history in broad strokes, is the frontier. Opportunity lies "out there" at the frontier, which is both spatial and temporal. Four such frontiers can be identified: *agricultural, urban-manufacturing, metropolitan,* and *rurban-cybernetic.* The original and best-known frontier is *agricultural. Open space* and *good soil* are keywords of the myth, but they also happen to describe, by and large accurately, what are there. The second frontier, *urban-manufacturing,* boasts innovative procedures in extracting and transforming nature—nature not, however, as open space and land but as coal and iron, steam and electricity. An archetypal frontiersman of this stage is a technically minded tinkerer and inventor like Thomas Edison rather than a hardy farmer or rancher. The third frontier, *metropolitan,* is a phenomenon of automated mobility, in particular the car. It too brings people to the edge of nature, but only if they are scientifically trained and can probe the mysteries of the physical-biological universe. And lastly, today America boasts the *rurban-cybernetic* frontier.[24] Nature at this frontier is remote indeed from the tangible things one encounters in ordinary life. It has become an abstract language encoded in softwares that have vastly increased the human ability to process and transmit information.

When the frontier was land, immigrants could move to it and immediately feel that they were at the "cutting edge" of forces that made for progress. When the frontier was urban-manufacturing, immigrants of peasant background found that they had to adopt a different way of life, including a different way of thinking, to establish a foothold at the cutting edge. As the frontier continued to shift, first to metropolitan, then to rurban-cybernetic, the lead to it became so demanding that new immigrants, unless they were of an educated middle class, felt discouraged by the distance of the goal and the effort that must be made to reach it. This was especially the case with the flood of poor Hispanic immigrants that entered the southwestern part of the United States beginning in the 1970s. Another class of Americans, those of

African descent, also felt discouraged in this post-civil-rights period. They are, after all, not new to the country. On the contrary, they and their forebears have lived on this land longer than any other ethnic group, the only exception being, of course, the American Indians. Yet they as a people remain relatively poor; many are stranded in the life and work ways of the earlier frontiers.

Discontent also grew among white Americans, as more and more of them struggled through college with degrees in the nontechnical fields to find themselves, upon graduation, in low-paid jobs that were unrelated to their academic pursuits. Moreover, even if they found jobs that drew on their credentials, such as teaching in colleges and universities, they received salaries well below those of their scientific colleagues. The resentment was not merely directed at the income disparity. It was also directed at the scientists' presumption that they, and they almost alone, were the pioneers of true knowledge. In the United States the scientists' point of view is so widely accepted that to the general public the expression *frontiers of knowledge* has come to mean primarily, if not only, scientific knowledge.

Rebellion among young whites would have been just a storm in the campus teacup but for their ability to ground their sense of injustice on the undeniable injustices suffered by African Americans. The glorious civil-rights movement, which sought to right a historical wrong and was aimed to benefit African Americans, was extended to cover all sorts of disadvantaged groups: other colored peoples, women, and homosexuals. However, the protests led by campus intellectuals, almost exclusively those with liberal arts degrees, have taken an unexpected turn. Rather than demanding equal opportunity to rise to the top of scientific and financial establishments, they have solved (quixotically) all injustices through the simple expedient of denying that there ever was a top in things that matter, such as ways of living and forms of knowledge. They deconstruct mountains so that all look like molehills. Anxious individuals and groups are assured that their particular molehill—their own pile of customs and beliefs—is all that they really need to reach the full potential of their humanity. Understandably, this profoundly conservative viewpoint is consoling to oppressed peoples who already feel discouraged by the many hurdles that they must overcome to reach the top, where the privileged sit.

Contradictions abound in a solution that is more sleight of hand than real. Unless one is drugged by rhetoric, of which there is plenty, it is easy to see that the modern world is so thoroughly penetrated by the workings and products of scientific culture that even the effort to organize the sacred corn dance, or, for that matter, protest movements against the hegemony of science depends on having the use of electric power, the Xerox machine, the telephone if not the fax. Moreover, disadvantaged minorities who try to resurrect ancestral ways may find that instead of gaining real confidence and dignity, they have lost the little they had by becoming "museum pieces" for the enjoyment of white Americans on safari. The oppressed have always had to offer their labor for sale. Mainline society, having gone hi-tech, doesn't seem to need their sinew and sweat any more, so in order to survive they must sell themselves— their heirlooms, their sacred rites, their own grandmothers decked out in native finery.

Awareness and Anxiety

Awareness is an instrument of survival: it keys an animal into the things that are good (food and mate) and into the things that are bad (pitfalls and natural enemies). In a sense, then, even animals may be said to have nibbled at the tree of knowledge of good and evil. Their punishment is a life that is never free from anxiety. With humans, greater awareness brings ever greater re-wards, but at a cost that we may not always be willing to pay.[25] How far and how clearly do we really wish to see? Knowledge is good, but is it always unambiguously good? Even in the modern West, a culture that has actively promoted knowledge, it might seem at times better not to have opened cer-tain doors. A recurrent theme in fairy tales is the door that must not be opened. When a child disobeys, she finds to her horror a world of blood, slaughter, and death.[26] Earlier I noted the existence of a universal unease with construction—with the building of anything that is more than necessary shel-ter. The feeling was that the gods would be offended by our presumption. Now, with greater knowledge, we won't put it quite that way. We are not afraid of offending the gods. We can, however, offend our own conscience, be deeply disturbed by the awareness that building up and tearing down are coordinate processes, that one can't have the one without the other, beginning with one's own body, which cannot be built up and maintained without tearing down, chewing up, and digesting plants and other living things—without the "blood, slaughter, and death" that the child in the fairy tale discovered behind the forbidden door. And if this is so true of construction of a mere biological body, what level of destruction must precede the construction of cultural works—villages and fields, towns and cities, whole civilizations? Who must suffer, what must be destroyed, so that we may progress?

In human relationship too an increase in awareness and knowledge, for all the rewards they bring, can raise dilemmas and challenges that become new sources of unease. Consider the understanding of American history in schools and colleges. Old texts used to boast simple-mindedly about the white settlers' victories and achievements. In the newer texts these are balanced by gory tales of misdeed: white pioneers, once cardboard heroes, now emerge as complex human beings, blends of good and evil. Without doubt, such a more complex picture represents progress. However, the treatment of Native Americans, African Americans, and nonwhite immigrants has not made similar progress. It remains one-dimensional, only one of a different kind, emphasizing not a people's deficiencies, as might have happened in the past, but their virtues. The educational establishment, in its anxiety to compensate for past neglect and unjust portrayals, has unwittingly brought about an unfortunate juxtaposition: a history of whites that is somber and adult along-side a history of nonwhites that can seem, by comparison, childishly inspirational. And so disrespect is once again shown to the oppressed even though respect is intended.

My final point concerns anxiety that is a consequence of our knowing so much about what is happening in other parts of the world. For the first time in human history we live, as the saying goes, in a global community, a global village. This adage is misleading, however, for the following reason. In a real village I can expect help from someone I have helped. The global village doesn't work that way. In the global village I am aware of the ills and misfortunes of people everywhere, and I am made to feel that I should sympathize and extend assistance, which I struggle to do with my wholly inadequate means, knowing at the same time that when I, personally, am in trouble I cannot expect strangers on the other side of the globe even to know, much less to help. Do I really want this lacerating stretching of sympathy and obligation that moral progress has made exigent and technological progress has made possible? To put it another way, awareness is good if it leads to my demanding more rights. But my demand for rights is empty unless there are people who take it upon themselves to respond. I, in turn, must consider other people's demands. And so I need to ask myself: ever ready as I am to extend my rights, am I as ready to extend my obligations?

Of course, the marvels of modern communication expose me not only to the miseries of the world but also to their wonders. The wealth of the globe is now at my disposal. Educators want students to be plugged into the world-wide Internet so that with a few clicks of the mouse, they can call up the treasures of Tutankhamen, the autographed scores of Bach's cantatas, the detailed design of an F-16, the teachings of Buddha, Confucius, or Stephen Hawking. Progress puts one in a dazzling emporium of knowledge. The result is what? Greedy consumerism of information leading to unprecedented superficiality? Dizziness, a sort of nausea of disorientation? The loss of attentiveness—the ability to immerse oneself for a long time in a particular branch of knowledge, tradition, or religious faith? Progress seems to consign us to a perpetual state of anxiety by offering us a giant menu for every conceivable need and desire; and yet to choose wisely, all we have to work with is a computer—the one in our head—that is now some 50,000 years old.

Notes

1. Martin Mayer, "The Closet Conservatives," *American Scholar*, spring 1977, 230.

2. A. T. Jersild and F. B. Holmes, *Children's Fears* (New York: Bureau of Publications, Teachers College, Columbia University, 1935), 118, 124.

3. This thesis is simply and classically stated by Mircea Eliade in *The Sacred and the Profane: The Nature of Religion* (New York: Harper Torchbooks, 1959).

4. Arthur Waley, *The Analects of Confucius* (New York: Vintage Books, 1938), 94.

5. Benjamin Schwartz, *The World of Thought in Ancient China* (Cambridge: Harvard UP, 1985), 148–49.

6. Gillian Gillison, "Images of Nature in Gimi Thought," in *Nature, Culture, and Gender*, ed. Carol P. MacCormack and Marilyn Strathern (Cambridge: Cambridge UP, 1980), 143–73; Mary Douglas, "The Lele of Kasai," in *African Worlds*, ed. Daryll Forde (London: Oxford UP, 1963), 1–26.

7. Kenneth R. Olwig, "Recovering the Substantive Nature of Landscape," *Annals of the Association of American Geographers* 86, no. 4 (1996): 630–53.

8. John Calvin, forty-ninth sermon, 30 July 1555, in *The Sermons of John Calvin upon Deuteronomie*, trans. Arthur Golding (London: Middleton, 1583), 295. See also Michel Foucault, *Madness and Civilization: A History of Insanity in the Age of Reason* (New York: Vintage Books, 1973), 56.

9. On the skyward reach of ancient Greek civilization see Vincent Scully, *The Earth, The Temple, and the Gods: Greek Sacred Architecture* (New Haven: Yale UP, 1962).

10. T. K. Chêng, *Archaeology in Ancient China: Shang China* (Toronto: U of Toronto P, 1960), 2:53–55.

11. Matthew Luckiesh, *Artificial Light: Its Influence on Civilization* (New York: Century, 1920), 158. See also Yi-Fu Tuan, "The City: Its Distance from Nature," *Geographical Review* 68, no. 1 (1978): 1–12.

12. For a well-illustrated study of the fascination with luxury ships sunk in "primordial" ocean, see Robert D. Ballard and Rich Archbold, with paintings by Ken Marschall, *Lost Liners* (New York: Hyperion Books, 1997).

13. Stanley Kauffmann, review of *Titanic*, directed by James Cameron, *New Republic*, 5 and 12 January 1998, 24–25.

14. Martha C. Nussbaum, *The Fragility of Goodness: Luck and Ethics in Greek Tragedy and Philosophy* (Cambridge: Cambridge UP, 1986), 1–21.

15. Alan Macfarlane, *The Family Life of Ralph Josselin: A Seventeenth-Century Clergyman* (Cambridge: Cambridge UP, 1970), 171.

16. Robert D. Sack, "A Sketch of a Geographic Theory of Morality," *Annals of the Association of American Geographers* 89, no. 1 (1999): 26–44.

17. I am influenced here by Karl Jaspers's postulation of an Axial Period (see *The Origin and Goal of History* [New Haven: Yale UP, 1953], 1–21).

18. Lewis Hyde, *The Gift: Imagination and the Erotic Life of Property* (New York: Vintage Books, 1983), 11–24.

19. Whether it is proper to eat meat used to be a fringe question in Western moral discourse. Since the 1970s it has moved closer and closer to the center. For a recent example of the debate see J. M. Coetzee with Marjorie Garber, Peter Singer, Wendy Doniger, and Barbara Smuts, *The Lives of Animals* (Princeton: Princeton UP, 1999).

20. C. S. Lewis, *The Discarded Image* (Cambridge: Cambridge UP, 1964); Jamie James, *The Music of the Spheres* (New York: Copernicus, 1995).

21. Blaise Pascal, *Pensées* (New York: Dutton, 1958), no. 206.

22. Ray Jayawardhana, "Tracing the Milky Way's Rough-and-Tumble Youth," *Science* 258 (27 November 1992): 1439; G. T. Sussman and Jack Wisdom, "Chaotic Evolution of the Solar System," ibid. 257 (3 July 1992): 56–62.

23. Steven Weinberg, *Dreams of a Final Theory: The Scientist's Search for the Ultimate Laws of Nature* (New York: Vintage Books, 1994), 164–65.

24. Daniel J. Elazar, *The American Mosaic: The Impact of Space, Time, and Culture on American Politics* (Boulder: Westview Press, 1994), 73–98.

25. Ralph Dahrendorf, *Life Chances* (Chicago: U of Chicago P, 1979).

26. "Fitcher's Bird," in *The Complete Grimm's Fairy Tales* (New York: Pantheon, 1972), 217; Roger Shattuck, *Forbidden Knowledge: From Prometheus to Pornography* (New York: St. Martin's Press, 1996).

Chapter 5

Perfectibility and Democratic Place-Making

J. Nicholas Entrikin

The goal of the humanities as liberal education has been described by the political philosopher Michael Oakeshott as inviting one to "disentangle oneself, for a time, from the urgencies of the here and now and to listen to the conversation in which human beings forever seek to understand themselves."[1] For Yi-Fu Tuan, such learning offers the opportunity for moral progress measured in our expanded abilities to see the outcomes of our actions as related not only to present circumstances but also to our knowledge of the experiences of others in different times and places. This conversation with temporally and spatially distant others offers self-knowledge that creates a "personal museum," which gives one's own life and world a sense of progressive enlargement. "After all," Tuan writes in *The Good Life,* "the well known aim of liberal education is to transform one limited self into a rich concourse of selves, one narrow world of direct experience into many worlds, a creature of one time into a seasoned time-traveler. It is the practical aim of liberal education to enable men and women to lead the good life."[2]

Tuan has recognized the fundamental relatedness of self, community, and place and the tensions that these relationships create in a world in which the goals and ideals of social harmony and wholeness compete with the disintegrating and anomic conditions of modern life. His work offers a complex portrait of modern human psychology in which ambivalence and ambiguity mix with passionate attachment and certainty. For Tuan, the discrepancy between the mind's search for order and symmetry and the fluid, changeable character of life offers a fundamental tension that underlies human attempts to create worlds out of nature.[3]

Human landscapes reflect this tension. According to Tuan, building allows humans a means of giving aesthetic and moral order to their worlds, and "none surpasses architecture in its power to present us with enveloping worlds that are also emblems of perfection. In building not only for community but also for splendor, architecture is or can be a moral 'science.'" In democratic societies these concerns of community and splendor come into conflict: "The word *splendor* is out of fashion, as is perhaps *vision:* both suggest a sort of hollow grandiosity at odds with the low-profile democratic ideals of our time."[4]

This conflict in the realm of architecture may be understood as an instance of a more general tension between perfectibility and democratic placemaking. The metaphysical concept of perfection or perfectibility—literally, the capacity to make perfect—when used in association with making places implies either the hyperbole of consumer culture or a potentially dangerous absolutism. In this latter sense the wholeness and organic quality of such places mirrors what has been described by Hannah Arendt as a totalitarian ideology in which individuals lose their unique identity through being welded together into larger social units, conceived as organic and guided by some larger force of nature or history.[5] Such "natural" social units have their "natural" places, where the link between community and place is characterized by necessity rather than contingency.

In the totalitarian state, a deformed public space hides or obliterates private space, but in democracies these spaces exist in constant relation, both harmonious and contentious. The source of such conflict is generally found in the mismatch between individual self-interest and the common good, or in the fragmentation of the political community into groups whose divergent goals undermine the hope of working toward the common good. Indeed, the political theorist Michael Walzer has identified these as the two primary centrifugal tendencies in modern American life: "One breaks loose whole groups from a presumptively common center; the other sends individuals flying off."[6] The common good may thus be conceived as the balancing of these often conflicting interests in a just and fair manner either through a more abstract and generalized version of the common good that encompasses the competing views or through an agreed-upon procedure for deciding among alternatives. The balancing of these different interests in a just and fair manner may work at cross-purposes with the desire for perfection. Tuan notes this conflict in *Passing Strange and Wonderful* when he states that "democracy, with its vocation to accommodate multiple—even conflicting—voices, would seem to rule out the possibility of creating anything coherent, much less lucidly harmonious." In Tuan's view, a democratic landscape aesthetics would give the appearance of fairness and equity, the essence of which he sees captured in the contrast between St. John de Crèvecoeur's description of late-eighteenth-century American cultural landscapes as offering a "pleasing uniformity of decent competence" and the splendor of Pierre L'Enfant's plan for the new nation's capital.[7]

Place-making within democratic societies would seem, then, to eschew the ethics or aesthetics of perfectionism for a more consensual model, one less metaphysical and more political. However, perfectionism has been and continues to be an element of modern democratic culture, especially in the United States. It is less evident as an American aesthetic ideal than it is as a moral argument. As a moral argument it is found in political theory in debates about civic virtue between supporters of republican and liberal models of democracy. A central concern in such debates relates to state neutrality and the extent to which a state may either passively or actively support a particular conception of the good life.[8]

More than 150 years ago Alexis de Tocqueville recognized the cultural ideal of perfectibility in a then young and evolving American society, writing in *Democracy in America* that Americans "have all a lively faith in the perfectibility of man, they judge that the diffusion of knowledge must necessarily be advantageous, and the consequences of ignorance fatal; they all consider society as a body in a state of improvement, humanity as a changing scene, in which nothing is, or ought to be, permanent; and they admit that what appears to them today to be good, may be superseded by something better tomorrow."[9]

Through Tocqueville and others perfectionism and the metaphysics of moral progress in American culture has been closely associated with the ideal of liberal education. This connection has ancient roots and has been modified to fit modern sensibilities. The goal of education in both its ancient and modern forms has been to produce critically thinking, cosmopolitan citizens who would maintain a moral orientation or compass in a world of constant flux. This model of learning should not to be confused with educational programs for a liberal political regime but might be viewed instead as "regime non-specific."[10] In classical arguments, liberal education was a means toward moral perfectionism, conceived as a teleological model of moral growth and progress. Modern proponents have adapted the classical and religious ideals of perfectionism to modern democratic societies by disconnecting perfectionism from its teleological roots. The relation of self and community is rephrased in a way that replaces the functionalist vocabulary of parts and wholes with that of process and becoming.

Perfectibility

The vocabulary of virtues in modern life has undergone what might be described as a semantic free fall. Their original lofty meanings seem to be confined to the pulpits or to the texts of moral philosophers. Terms such as *perfect* and *perfection* are frequently heard and read in everyday life, but primarily in reference to a world of quantifiable standards. One hears of perfect scores in both scholastic and sporting events and thus of perfect performances. The most dramatic examples of the prosaic quality of the vocabulary of virtues are in consumption and advertising, from the perfect car to the perfect body or the perfect look. Perfectionism thus becomes the match of product design and individual desire.

The only connection between these everyday meanings associated with modern consumption and performance and an original religious meaning related to divine perfection is the recognition of an ideal against which all things are measured. The philosopher H. P. Owen referred to the two overlapping meanings of *perfection* in Greek thought, associated with the terms *teleios* (perfect) and *telos* (goal or end).[11] The first refers to some sense of "wholeness, completeness, and integrity," and the second refers to achieving a goal in the Aristotelian sense of realizing a specific form or a natural place. Together, they suggest that an object or being is complete in fulfilling its nature.

In Judeo-Christian thought and in other religions perfection is manifested in the divine, but this view is not universal. Many of the Greek gods were less perfect than their earthbound believers but nonetheless had the distinct advantage of immortality. Moral teachings set within religious frameworks frequently manifest this perfectionist theme. Such themes transfer to humanist, secular thought in the appreciation of the goal of living a life that satisfies an ideal of humanity, such as human nature, true being, authenticity, and so on.

Perfectionism was also part of ancient and medieval philosophy. For example, Plato's *Republic* describes the ideal citizen and state, and the Stoics linked perfection to the control of the passions through reason, which reflected that portion of the divine that resides in humans.[12] Christian philosophers such as St. Thomas Aquinas distinguished between the absolute and infinite perfection of God and the more limited perfectibility of humans, and St. Augustine used perfectibility as a distinguishing marker that separates the City of God from the City of Man. Each recognized a completeness that comes only with the divine.[13]

Such completeness lifts one burden while adding another. The transcendent spirit of Judeo-Christian religious belief helped to lift its believers out of the darker shadows and anxieties created by the practices of magic and witchcraft but in doing so added the burden of an "insistent perfectionism that would have been annihilating but for the continuing existence of laws and regulations, customs and practices, dispensations, and indulgences."[14] Culture thus presents the tools for translating the ideals of divine perfection into the decidedly imperfect world of human practices.

Moral perfectionism as a modern theme draws from this classical and religious foundation. It offers a teleology for the production of great individuals and a guide toward an authentic existence in the writings of Friedrich Nietzsche and Martin Heidegger. Nietzsche wrote that one's life gains greatest value in "living for the good of the rarest and most valuable specimens," and Heidegger's existentialism posed a self in search of authenticity.[15] Its more social manifestations are found in the writings of Marx, who emphasized the importance of cooperative, egalitarian communal relations, in which no individual may fully develop his or her potential unless all are able to do so.[16]

In contemporary American thought, moral perfectionism has been a central theme in Stanley Cavell's recovery of the nineteenth-century transcendentalists Ralph Waldo Emerson and Henry David Thoreau as modern moral philosophers.[17] Cavell and his colleague Hilary Putnam have examined the relationship between self-perfectionism and democracy. Both philosophers "explore democracy not in terms of its concrete rules and political institutions, but rather as an ethical ideal that is central to what we regard as the primal question, 'how to live?' "[18] Cavell's arguments especially seem to intersect at several points with those of Tuan as a consequence of their common exploration of the nature of the good life through reading, writing, and conversation. Cavell, however, directs his attention specifically at questions of democratic practice and self-realization, issues that are addressed by Tuan but in a less meliorist fashion. For Cavell, moral perfectionism is not only compatible with but also essential to democratic forms of life.[19]

Cavell on Perfectionism and Democracy

Cavell recognizes in Plato's *Republic* numerous themes of perfectionism, which are presented in discussions of moral education and conversation.[20] Such conversation concerns the model of an exemplary life and the process by which a trapped, enchained self is liberated through expanded moral awareness that moves beyond the self to the social. Plato's path toward perfectionism leads to aristocracy. It is a path of self-enlightenment for the few that form the aristocratic class. Cavell contrasts this Platonic aristocratic ideal with democratic perfectionism by generalizing this enlightenment process to the society as a whole: "a path from the idea of there being one (call him Socrates) who represents for each of us the height of the journey, to the idea of each of us being representative for each of us—an idea that is a threat as much as an opportunity."[21]

The concept of the self according to Cavell's description of perfectionism is one that is necessarily split, offering a dual conception divided in Heideggerian fashion between the inauthentic self and the self seeking authenticity. The path toward authenticity is not a purely individualist one, although it is one that each individual must go through. It is a path followed in relation to others, but first one's sense of self must be directed and clarified. One attains genuine selfhood through "maintaining a relationship to an exemplary other."[22]

In Plato's classical model of perfectionism the perfect exists in the pure realm of ideas. Emerson, as interpreted by Cavell, transformed this transcendental realm into a process. The self is a continual dual of the "attained" and the "unattainable"; each state is a final point and a beginning point. For Emerson and Cavell the self is always both attained and yet to be attained.[23]

The tools of this transformative process are thinking, reading, and conversation. Its political end is the development of self-reliant citizens in a democratic community. Such self-reliance provided a barrier to the pressures of conformity that are part of any collective but that Emerson saw as a threat to democratic community. Likeness of thought led to static moral judgments, or "moralisms," and adherence to a fixed set of moral principles. The critical spirit of democratic deliberation was thus undermined. For Cavell "the mission of Perfectionism generally, in a world of false (and false calls for) democracy, is the discovery of the possibility of democracy, which to exist has recurrently to be (re)discovered."[24]

In keeping with Tocqueville's concern for a strong sense of the individual in maintaining a stable democracy, Cavell insists on the essential Emersonian relation between critical self-realization and democracy. Such a relation entails "an independent, non-conformist, self-perfection . . . consistent with democracy's egalitarian concern for good and justice for all."[25] Cavell's perfectionism offers an internal means of self-critique for democratic societies "by providing the sort of caring, demanding self-improving individuals who can best guarantee that the institutions and practitioners of democracy will not rest content with the always imperfect justice and improvable good they provide."[26]

Democracy and Liberal Education in American Thought

In contemporary political theory moral perfectionism may be linked to those who concern themselves with the threats to democracy that come from conformity and leveling tendencies. Thus support is more likely to be found among defenders of certain republican, as opposed to liberal, conceptions of democratic community. For the republicans civic virtue rests at the core of genuine democratic community, and for the liberals such community is founded upon agreed-upon procedures for reaching fair and equitable judgments that protect the rights and autonomy of its members. This distinction represents a continuum along which various positions may be arrayed. For example, Richard Dagger seeks to mediate these polar positions through what he defines as republican liberalism: "Without ever expecting perfection, a republican liberal will aim to promote the civic virtues that enhance the individual's ability to lead a self-governed life as a cooperating member of a political society." "The republican-liberal citizen," writes Dagger, "is someone who respects individual rights, values autonomy, tolerates different opinions and beliefs, plays fair, cherishes civic memory, and takes an active part in the life of the community." Thus a republican-liberal education becomes one that promotes autonomy and civic virtue by connecting rights and responsibilities.[27]

According to the leading liberal theorist, John Rawls, the variants of the moral perfectionist argument come in two basic forms. One offers a teleological theory of society in which all institutions are directed toward the achievement of human excellence, even at the occasional cost of injustice. In its extreme such an argument offers a justification for slavery in classical Greek society based on the high level of achievement of Greek civilization. The second, more moderate form identifies excellence in achievement as one among several standards that the society may apply, and its third and weakest usage simply suggests that excellence in certain circumstances should be valued. All forms of perfectionism fail Rawls's test of social justice.[28]

Republican theorists pose a somewhat different set of questions. For example, Thomas Pangle shifts the emphasis from social institutions to the nature of the good and to civic virtue: "The great doubt classical republican theory poses for modern republican theory is this: has modern theory, in its successful attempt to clarify and satisfy the most basic legitimate demands of political life, obscured the clear view of human excellence that is required in order to shape a public life that reflects the whole of the common good?"[29]

Pangle responds affirmatively and identifies education as a central issue.[30] He argues that the modern university is caught between a scientific curriculum that offers a useful training for a modern, technologically sophisticated work force to compete in a competitive world economy and a humanities curriculum that has lost its way in the skepticism of postmodernism. The gap left by the absence of a tradition of critical liberal education contributes to a conformity of thought among teachers and their students that threatens to undermine democratic community.

Once again Tocqueville becomes the prescient voice describing the nature of this threat, which he saw as a public made docile by the conforming pressures of public opinion. In comparing aristocracy and democracy he noted two contradictory tendencies: the unequal conditions of aristocracy give no value to mass opinion, but in conditions of equality public opinion dominates. Tocqueville's prescription is an education in the classics, which offer lessons on the power of the individual to act effectively and thus to influence the course of events. Pangle concurs by endorsing a classics-based liberal-education model as a counter to the easy relativism of modern skepticism, which, he asserts, masquerades as democratic tolerance.[31]

My argument may appear headed toward the widening chasm between a great-books liberal-education model and a multiculturalist and postmodernist model. However, I would like to try a detour around this divide by considering briefly another conception of education and democracy through American pragmatism, most specifically in the thought of John Dewey. Dewey too was concerned about the excesses of both individualism and the conformity of mass society, but he has at the same time been cited as a direct intellectual ancestor to Richard Rorty, one of the primary philosophical architects of postmodernism. Dewey, like Emerson and Cavell, offered an ethics of self-realization that was essential for the achievement of desired social ends.

Dewey's ideas of modern democracy took careful account of science and its potential. The scientific spirit valued fair-mindedness, intellectual integrity, and the will to subordinate the interests of the individual for the good of the group. The method of science shared important moral traits with democracy, a valuing of innovation, openness, and cooperation and a devotion to progress.[32] The scientific spirit of inquiry became a cornerstone of Dewey's philosophy of education.[33]

Dewey is probably best known for emphasizing the social character of individualism against the American ethos of individualism that prevailed at the turn of the century. However, he also saw the dangers of a subject or self overwhelmed by the social, especially with the growing scale of modern societies, in which individuals feel increasingly powerless. Such concerns contributed to his emphasis on community, organized around communication and cooperation and rooted in place, the model that most geographers recognize through the innovative urban research of one of Dewey's students, the sociologist Robert Park. For Dewey, social relations were best built and maintained at the scale of the local community: "Democracy must begin at home, and its home is the neighborly community."[34]

According to Dewey, democracy as a way of life must be distinguished from democracy as a set of institutions.[35] As a way of life it originates in the local, in the home, the community, the school, and the factory. For Dewey, the ability to move from the local to the more abstract concern with the human community required a model of learning that had history and geography at its core. For example, Dewey observed that

> the earth as the home of man is humanizing and unified; the earth viewed as a miscellany of facts is scattering and imaginatively inert. Geography is a topic that originally appeals to imagination—even to the romantic imagination. . . . while local or home geography is the natural starting point in the reconstructive development of the natural environment, it is an intellectual starting point for moving out into the unknown and not an end in itself. . . . when the familiar fences that mark the limits of the village proprietors are signs that introduce an understanding of the boundaries of great nations, even fences are lighted with meaning.[36]

Dewey's philosophy of education emphasized self-realization of one's talents toward the goal of contributing to the well-being of the community. His model of liberal education was not one of great books, however, as was made evident in his public debate with Robert M. Hutchins during the 1930s. Dewey sought to replace the traditional metaphysical core of the liberal-education project with themes of liberal democratic theory.

The Dewey-Hutchins debate has been partially replayed in a review by Richard Rorty of Allan Bloom's *Closing of the American Mind*, in which Rorty argues that

> what matters to us [pragmatist] "intellectuals" as opposed to the [metaphysical] "philosopher," is the imaginativeness and openness of discourse, not proximity to something lying behind discourse. Both Platonists and Deweyans take Socrates as their hero. For Plato, the life of Socrates did not make sense unless there was something like the Idea of Good at the end of the dialectical road. For Dewey, the life of Socrates made sense as a symbol of a life of openness and curiosity. It was an experimental life—the sort of life that is encouraged by, and in turn encourages, the American democratic experiment.[37]

Republican theorists have followed Dewey's commitment to community, if not his pedagogical theory, in their attempts to adapt the liberal-education model of learning to complex modern societies. For example, Michael Sandel outlines a framework for a liberal-education model that supports the development of civic virtue. He describes the republican model as one having a concept of liberty that "depends on sharing in self-government" in a way that requires more commitment than simply the individual's choice of community participation as a means toward personal ends. Sharing in self-rule means, for Sandel, a deliberation with fellow citizens about a common good for the community. Such deliberation requires "a knowledge of public affairs and also a sense of belonging, a concern for the whole, a moral bond with the community whose fate is at stake. To share in self rule therefore requires that citizens possess, or come to acquire, certain qualities of character, certain civic virtues."[38]

Sandel's model of the self differs, however, from that of the classical republican tradition, which viewed self-government "as an activity rooted in a particular place, carried out by citizens loyal to that place and the way of life it embodies."[39] Self-government today requires a politics that develops at many different spatial scales, from that of the neighborhood to that of the world community. This condition puts added pressures on political actors. They must learn to act, writes Sandel, as "multiply-situated selves," who must work within a context of overlapping and competing claims of obligation and duty and of multiple loyalties.[40]

Virtues may become vices when the strain of this multiplicity becomes too much. Sandel notes two especially troublesome outcomes of such stress, fundamentalism and the possibility of "storyless selves," that is, fragmented selves with no integrating and guiding narrative. The loss of the ability to create coherent narratives of self leads to a sense of hopelessness and disempowerment of self. Both constitute attempts to hide the self from view, the first through its complete submergence into a social group and the second through its disappearance into chaos.

Conclusion: Democratic Place-Making

If Robert Sack is correct in arguing that place and self are mutually constitutive, then the means of creating the ideal self for sustaining the project of democracy have parallels in place-making.[41] In constructing places we seek to have them match our projects and ideals, both individual and collective. In democracies such ideals include the desire to build places that promote social justice, tolerance, and inclusion and that offer public gathering spaces or places reflecting collective values about community and the natural environment. At the same time, we also recognize that dysfunctional, in this case undemocratic, places exist, such as those that promote injustice or that unfairly exclude.[42] Viewed as a form of life and as a process, democracy involves in part the making, unmaking, and remaking of places.

In *Passing Strange and Wonderful* Tuan offers aesthetic criteria for judging these and other democratic constructions, aesthetic criteria that have been transformed into moral standards.[43] These are that great human works should demonstrate unity or wholeness that encompasses diversity and should be able to develop or grow without losing their original character. These criteria seem to take us back to *teleios* and *telos*, but without the finality of the first or the teleology of the second. The problem, however, is that in the complex democracies of modern societies consensus on such matters often seems beyond reach. As Tuan has observed, true lovers of American democracy must have a well-developed tolerance for contradiction.[44]

It is this sense of contradiction that casts shadows over all human attempts to constitute harmonious wholes that resolve the fundamental tensions between self and group, nature and culture. Harmony and balance provide a powerfully attractive yet elusive goal. The implied sense of equilibrium, even a so-called dynamic equilibrium, is a human construction poorly suited to describe natural processes or the flow of human experience. In its purist expression as a cultural ideal such changeless existence is part of the description of paradise as the perfect place.[45]

This ideal is found in political discourse on democracy in Rousseau, who described the perfection of an idealized Geneva. It is a place of enlightened, educated citizens. However, a secure political identity required a relatively impermeable and isolated community. In his analysis of Rousseau's *Letter to D'Alembert* David Steiner writes: "Where there are no physical walls, Rousseau suggested, education must construct them in the mind."[46] Democratic education creates citizens who are capable of forming rules that allow self-governance and promote civic virtue, but rules that require containment both of self and place and relatively stable or unchanging ways of life.[47]

A similar sense of wholeness that allows for an open and harmonious relation of self, community, and nature has been an important strand in American culture from the transcendentalists through the pragmatists. For Emerson's fellow transcendentalist Henry David Thoreau and for the prag-

J. Nicholas Entrikin

matist John Dewey a goal of education and self-reflection was to make the self at home in the world, as part of nature and human community. They recognized, however, that change must be a fundamental feature of such understanding. Thus, each characterized nature and experience as an open process or emergent flow that the human imagination seeks to understand and order through concepts.

For Emerson and Dewey the relation of humans to nature was "always possibility, often celebration, frequent mishap, and never absolute certitude."[48] Democratic forms of life offer an added element of uncertainty and instability that reflects the openness of public discourse about community ideals and values and the related possibility of abrupt changes in the goals of collective projects. Unlike the paradises of the popular imagination and the world's great religions, and unlike Rousseau's Geneva, in which permanence is part of their perfection, the places of democratic societies—homes, communities, nations—must always be in process, constructions to be maintained and repaired.[49] Democratic places mirror the duality of the self described by Cavell as the attained and the attainable.

This duality at the center of Cavell's moral perfectionism presents an ideal model of the process of self-awareness, which he sees as necessary to building democratic citizens and maintaining democratic community. It is a self-awareness that comes from making oneself intelligible: "Perfectionism's emphasis on culture or cultivation is, to my mind, to be understood in connection with this search for intelligibility, or say this search for direction in what seems a scene of moral chaos, the scene of the dark place in which one has lost one's way. . . . perfectionism's obsession with education expresses its focus on finding one's way rather than on getting oneself or another to take the way."[50]

Tuan has sought a similar goal of finding one's way and, in a spirit similar to that of American philosophy, offers an optimistic assessment of the illuminating and transformative powers of education.[51] His is not a naive optimism, for the world that he describes is always one of opposing forces of light and dark, good and evil. His recognition of these enduring oppositions does not, however, lead him to skepticism and despair, but rather to a cautious optimism about moral progress and its possibility within often chaotic and fragile democratic forms of life. Underlying this optimism is the ideal of individuals finding their way through an examined life and the sympathetic assessment of the continual attempts of individuals and communities to perfect necessarily imperfect worlds.

Notes

1. Michael Oakeshott, "The Place of Learning," in *The Voice of Liberal Learning: Michael Oakeshott on Education,* ed. Timothy Fuller (New Haven: Yale UP, 1989), 41, cited in René Vincente Arcilla, *For The Love of Perfection: Richard Rorty and Liberal Education* (New York: Routledge, 1995), 3.

2. Yi-Fu Tuan, *The Good Life* (Madison: U of Wisconsin P, 1986), 161.

3. Yi-Fu Tuan, "Ambiguity in Attitudes toward Environment," *Annals of the Association of American Geographers* 63 (1973): 411–23.

4. Yi-Fu Tuan, "Moral Ambiguity in Architecture," *Landscape* 27 (1983): 17.

5. Hannah Arendt, *The Origins of Totalitarianism* (London: André Deutsch, 1986), 446, cited in Kimberly Hutchings, *Kant, Critique, and Politics* (London: Routledge, 1996), 83.

6. Michael Walzer, *On Toleration* (New Haven: Yale UP, 1997), 93.

7. Yi-Fu Tuan, *Passing Strange and Wonderful: Aesthetics, Nature, and Culture* (Washington, D.C.: Island Press, 1993), 200, 204.

8. George Sher, *Beyond Neutrality: Perfectionism and Politics* (Cambridge: Cambridge UP, 1997).

9. Alexis de Tocqueville, *Democracy in America* (1835; reprint, New York: Vintage Books, 1990), 393; also cited in Stanley Cavell, *Conditions Handsome and Unhandsome: The Constitution of Emersonian Perfectionism,* The Carus Lectures (Chicago: U of Chicago P, 1990), 15.

10. Richard Flathman, "Liberal versus Civic Republican, Democratic, and Other Vocational Educations: Liberalism and Institutionalized Education," *Political Theory* 24 (1996): 24.

11. H. P. Owen, "Perfection," in *The Encyclopedia of Philosophy,* ed. Paul Edwards, vols. 5–6 (New York: Macmillan, 1974), 87.

12. Martha Nussbaum, "Kant and Stoic Cosmopolitanism," *Journal of Political Philosophy* 5 (1997): 1–25.

13. Owen, "Perfection."

14. Yi-Fu Tuan, *Escapism* (Baltimore: Johns Hopkins UP, 1998), 187.

15. See John Rawls, *A Theory of Justice* (Cambridge: Harvard UP, 1971), 325 n. 51.

16. Thomas Hurka, *Perfectionism* (New York: Oxford UP, 1993), 177.

17. Cavell, *Conditions Handsome and Unhandsome,* 1.

18. Richard Shusterman, "Putnam and Cavell on the Ethics of Democracy," *Political Theory* 25 (1997): 193.

19. Cavell, *Conditions Handsome and Unhandsome*, 28; idem, *Philosophical Passages: Wittgenstein, Emerson, Austin, Derrida* (Cambridge: Blackwell, 1995), 56; Shusterman, "Putnam and Cavell on the Ethics of Democracy," 202.

20. Cavell, *Conditions Handsome and Unhandsome*, 6–8.

21. Ibid., 9.

22. Stephen Mulhall, *The Cavell Reader* (Cambridge: Blackwell, 1996), 354.

23. Cavell, *Conditions Handsome and Unhandsome*, 12–13.

24. Ibid., 16–17.

25. Shusterman, "Putnam and Cavell on the Ethics of Democracy," 202.

26. Ibid.

27. Richard Dagger, *Civic Virtues: Rights, Citizenship, and Republican Liberalism* (New York: Oxford UP, 1997), 195–96, 131.

28. Rawls, *Theory of Justice*, 325–32.

29. Thomas Pangle, *The Ennobling of Democracy: The Challenge of the Postmodern Age* (Baltimore: Johns Hopkins UP, 1992), 9.

30. Ibid., 163.

31. Ibid., 216.

32. Ibid., 64.

33. John Dewey, *Democracy and Education: An Introduction to the Philosophy of Education* (1916; reprint, New York: Free Press, 1944).

34. John Dewey, *The Public and Its Problems* (Denver: Alan Swallow, 1927), 213.

35. Robert B. Westbrook, *John Dewey and American Democracy* (Ithaca: Cornell UP, 1991).

36. Dewey, *Democracy and Education*, 212.

37. Richard Rorty, "That Old-Time Philosophy," *New Republic*, April 4, 1988, 31, cited in Arcilla, *For the Love of Perfection*, 106.

38. Michael Sandel, *Democracy's Discontent: America in Search of a Public Philosophy* (Cambridge: Harvard UP, 1996), 5–6.

39. Ibid., 350.

40. Ibid.

41. Robert Sack, *Homo Geographicus* (Baltimore: Johns Hopkins UP, 1997).

42. Ibid.

43. *Passing Strange and Wonderful*, 208.

44. Yi-Fu Tuan, *Cosmos and Hearth: A Cosmopolite's Viewpoint* (Minneapolis: U of Minnesota P, 1996).

45. Yi-Fu Tuan with Steven Hoelscher, "Disneyland: Its Place in World Culture," in *Designing Disney's Theme Parks: The Architecture of Reassurance*, ed. Karal Ann Marling (New York: Flammarion, 1997). See also Yi-Fu Tuan, *Dominance and Affection: The Making of Pets* (New Haven: Yale UP, 1984).

46. David M. Steiner, *Rethinking Democratic Education: The Politics of Reform* (Baltimore: Johns Hopkins UP, 1994), 67.

47. Ibid., 108–13.

48. John J. McDermott, *Streams of Experience: Reflections on the History and Philosophy of American Culture* (Amherst: U of Massachusetts P, 1986), 35.

49. Tuan, *Escapism*, 202–3.

50. Cavell, *Conditions Handsome and Unhandsome*, xxxii.

51. Yi-Fu Tuan, *A Life of Learning*, Charles Homer Haskins Lecture, American Council of Learned Societies Occasional Paper No. 42 (New York: American Council of Learned Societies, 1998).

Chapter 6

Geographical Progress toward the Real and the Good

Robert David Sack

The progress I discuss here is not simply change, it is change in the direction of a goal. But not any old goal will do, for that makes progress relative. Rather it must be a goal that matters and endures, that is distant and demanding, and that is not ultimately attainable. It must be a goal that acts as a lure, drawing us in the right direction along a never-ending journey. If the goal were anything less, if it were attainable, it would then become unworthy as a grand ideal and would end the possibility of progress.

I assume that the most worthy yet unreachable goal is a heightened awareness of the real and the good. The two are intimately related. I believe that reality exists. We most definitely contribute to it—adding to, altering, and subtracting from reality—but we do not make it up. It is not simply a product of our imaginations. Reality exists, but it is infinitely complex, and our knowledge of it is always partial and provisional. Reality ultimately is ineffable.[1]

The good also is real. We undertake actions that are good or bad, and there are infinite ways of doing so. Our actions provide diverse manifestations of the good, and they contribute to the amount of goodness in the world, but we do not make up or invent what it is that makes these good. We may have provisional and partial senses of what the good is, but it too is infinitely complex and ultimately unknowable. The good, then, is real, but we also must realize that not all of reality is good.

Geography is intimately involved in this relationship between the real and the good. An essential quality of our lives is that we are place-makers. We create places—delimited and controlled areas of space that contain rules about what may or may not take place (or places-as-territories)—so that projects can occur. Virtually all of our undertakings require such places, and they range from the small scale of the room to the large scale of the nation-state. These places and the activities they support affect reality. Even though we influence only a small portion of the real as it exists in the vastness of the cosmos, the area we do affect—the earth—is for us a most important part of the real. Place affects reality in terms of adding, subtracting, and altering activities. Building a new city is such an alteration or an addition. So, too, is building a slave plantation or a concentration camp.

Not only does our place-making activity contribute to the real, it contributes to the good—or the bad. Homeless shelters, hospitals, and even universities, if they adhere to their ideals, can be thought of as good places; crack houses and abusive homes are not. The moral role of place is not confined to isolated events but undergirds entire systems. Slavery in the American South and the Holocaust in Nazi Germany were undergirded by complex systems of places. The American system of slavery required places in Africa in which new slaves were assembled to be transported to numerous and varied places on the North American continent, with the slave plantation a central component in the process. Similarly, Nazi Germany required a host of places and territorial controls to help it attain its goals, and concentration camps were central to its project of exterminating the Jews.

Place is connected to the real and the good in another way. As for the real, some places can make us more—or less—aware of reality and of how our geographical agency alters it. This is important because our degree of awareness of the real affects how we contribute to reality. If we do not know what our places and projects accomplish and how they affect other places and projects, then we may not be able to contribute to reality in the way we expect. And awareness also affects how we contribute to the good. As moral agents we must be responsible for our actions. This means that we must be aware of the consequences of our undertakings. Places that help us become more aware can increase our capacity to do the right thing, and places that diminish awareness can reduce this capacity. This relationship between awareness and the good is based on the assumption that not only does goodness require an awareness of reality but a good deal of evil results from a failure to understand what it is that we are doing and of the possible alternatives that might be available. This role of awareness is based on the well-known argument that if we knew that something was evil, we would not select it. That is, we do not do evil willingly; rather, evil occurs because of our inability to see clearly. Hence the belief that an open and educated society would be a good one. Evil in this respect can result from a lack of social opportunity or individual effort on our part to become more aware. A lack of awareness is an important cause of evil simply because it results in a narrowing of moral concern.

It is good, then, to be more rather than less aware so that we can under-stand the consequences of our actions, and it follows that places that expand and heighten our awareness are good places. They help us to do what is right. They point us in the right direction and assist us in moving toward the real and the good. Moving toward the real and the good is what I take to be the essential element of moral progress. My purpose here is to consider how geography helps us make progress toward the real and the good.

Saying that the real and the good exist may, in some minds, raise the specter of intolerance and absolutism. This is the case only if we believe that we can know the real and the good completely and certainly. But we cannot, for they are largely ineffable. Ineffability makes all of our knowledge provi-sional and our disposition open. A reality that is completely known no longer holds our attention or draws us to inquire. A good that seems determinate and attainable is no longer something to which we need to be open and for which we strive. Such a "good" is not good but rather a diminished set of proscriptions that create a yoke around our necks. The position I am taking here, of the real and the good that exist but that are ineffable, provides a middle ground between an absolutism on the one hand and a relativism on the other.

My point is that places that help us see the real and the good more clearly are intrinsically good places; they help us make real or *intrinsic geographical progress*, based on *intrinsic geographical judgments,* which are judgments in-formed by the good. These kinds of places can be very different from, and far rarer than, places that are judged to be good solely because they are effective instruments in attaining particular goals or in undertaking particular proj-ects. Evaluating places only in light of their instrumentality, that is, employ-ing *instrumental geographical judgments,* leads to the common-sense idea of *instrumental geographical progress*. Instrumental progress (and instrumental judgment) can be limited and relative. By itself, instrumental judgment and progress may not open up the possibility of real moral progress. By contrast, intrinsic progress transcends particular goals and projects and helps us to make progress in terms of seeing the real more clearly and of doing good.

This is an overview of what I have to say. I address these relationships in three stages. I first discuss the relationship between instrumental geographical judgments and progress and then the relationship between intrinsic geographical judgments and progress. Finally, I discuss the connections between instrumental and intrinsic, especially how excellence and aesthetics provide the possibility of a bridge between the two.

Instrumental Geographical Progress

Geography focuses on how we transform the earth and make it into a home. The countless projects involving our transformation of nature into culture and the creation of yet more culture require geographical tools or instruments that help us undertake this work. Among the most important of these is place—as an area of space we humans bound and control with rules about what may or may not take place. Projects require place. Political activities and power are enforced through place, whether it is the hunting-gathering territories of preliterate peoples, the estates and manors of feudalism, or the territorial units of modern nation-states. So, too, does economic activity as it takes place within market areas, along roads and railroads, and within factories, farms, and offices. Even though some projects need only one type of place, others happen in a variety of places. Still, there may be one or two that dominate. The primary task of raising families takes place in homes. The primary task of education takes place in schools. For athletics, the place is gymnasiums and sports arenas, and for relaxation it may be movies and amusement parks. Even the supposedly unbounded world of nature now takes place mostly, or most intensely, in nature preserves and wilderness areas, which are places we create in order to prevent us from transforming nature into culture; yet paradoxically, we often must intervene here to keep nature as we want it to be.

Note that our focus is on place but that places are constructed and maintained by us. We are the ones who bound the areas of space with rules about what may or may not take place, and we are the ones who change and transgress these rules. Place, in the sense I am using the term here, cannot exist without us, but we cannot exist without it. Place and self are relational. The place we create enables and empowers us by helping us organize reality. This means that it has an effect on our undertaking and projects in the same way as do other indispensable tools, such as language. Since this is a geographical discussion, the emphasis is on the effect and power of place as an instrument. Thus, I speak of places making this or that happen and of places drawing and mixing or weaving things together, but this should not obscure the fact that we humans are the ones who create places and initiate projects, just as we are the ones who have cumulatively created and speak languages. But we cannot communicate without language, and we cannot undertake projects without place. It is this essential causal property of place that I emphasize here. It must also be stressed that the singular term *place* often stands for a system of places and their flows and interactions through space.[2]

Place, then, not only helps create and sustain projects but contributes to the real. Some places are more important than others, but all places are equally real. Our reality can become richer and more complex as the number and types of places increase. In this sense, cultural reality, at least, was simpler for humans thirty thousand years ago, when the built environment contained only hearths, huts, and hunting grounds, and not also factories, offices, towns, cities, and nation-states. The places and practices we engage in can thus expand geographical reality by simply multiplying in numbers and variety. But the relationships between places and the real do not always move in that direction. Some types of place can become dominant and thus diminish the need for others, for example, when all large urban centers come to look alike, or when shopping malls replace the smaller, more varied types of stores, or when the expansion of cattle ranching diminishes the variety of natural habitats. So even though there may be an increase in number, these places can diminish the variety and complexity of reality. Thus, our actions as geographical agents affect reality: they can cause it to expand or contract.

But place and reality are also linked with regard to our awareness of the real. Some places help enhance our awareness of reality. I take this to be a primary mission of schools and universities. Other places seem to be indifferent, and some places help encourage us to discard reality for make-believe or fantasy. I take this to be the point of amusement parks and many tourist resorts. And some places diminish our awareness of reality. These include places that censor information and places that are secret.

To sum up, then, places are necessary instruments for projects. Places can increase or decrease the real, and some places can expand or contract our awareness of the real. Still, if we think only in terms of instrumental progress, then regardless of whether the outcome is an expansion or a contraction of the real or of our awareness of it, we will make instrumental progress as long as the place helps us do what we want. Instrumental geographical progress is, then, relative to the goals of a project: it leads to a concept of mores or custom rather than a concept of morality.

Intrinsic geographical progress, on the other hand, which is based on the idea of the real and the good, avoids this relativism. But before we consider this other form of progress, we need to explore further what makes place instrumentally useful and essential for undertaking projects.

The key is that place is an indispensable device that allows us to include some things and exclude others. The rules about what should or should not take place and the enforcement of these rules through the boundaries of place, clear a space for our activities. In clearing a space, place not only helps us include things but also draws and mixes them together. In general, places help draw together elements of nature and culture. Culture can itself be thought of as comprising social relations and meaning. So place helps to draw and mix together elements of nature, social relations, and meaning.

Consider the lecture hall. It draws some elements of the natural world, including light, heat, and ourselves as biological beings, yet it excludes a host of others, including rain, cold, and stray animals. It also draws together elements of social relations. It includes students and teachers, who also may be husbands and wives, friends and colleagues, rich and poor. But the room is not open to every type of social relation. It is not a place for the homeless, nor is it a place that corrects social deviancy. The elements of nature and social relations that are present here play supporting roles to the primary purpose of the place, namely, its focus on meaning. The mix of nature and social rela-tions supports the emphasis on meaning, or at least that is the intent. The effectiveness of the place depends on how well this mix works.

The mix can change, of course, as can the focus of the place. It can happen subtly, as a result of lowering the temperature a bit, or including more students, or changing the focus of the talk, or it can shift dramatically, so that it is no longer dominated by meaning but rather by social relations or nature. In the middle of a lecture a stranger might stumble in, clutching his chest, and ask for help. It would then become the social responsibility of those in the lecture hall to shift from meaning to the care of a stranger. Or if it were the middle of winter and the heating unit were to fail, causing the temperature to plummet, those in the hall would be in the grip of nature.

Different places establish different mixes for different projects. Schools and universities focus on meaning, offices and factories focus on particular social relations among workers and managers, and parks and wilderness areas focus on the elements of nature. These things that places include and draw together form the bases of our projects and are part of what makes place essential to their execution. A place works well if it draws the mix appropriate to the project, and it does not work well if its mix is not right. And when place works well we tend to become unaware of it, as it recedes into the background. But once something is wrong or out of place, then place becomes the object of our attention.

Instrumentally, then, we make geographical progress when place works well, that is, when the boundaries, rules, and mix of elements allow us to progress toward the goals of the project. This is all well and good as far as it goes. But we need also to judge the place and its project in a more general sense. Are they good? What is the point of calling something progress when it leads to bad or evil outcomes? With different rules and mixes, place helps perpetuate dysfunctional homes where children are beaten and abused. Place in the form of Nazi concentration camps helped eradicate the racially "impure," and places in the form of slave plantations were the mainstay of slavery in the American South. The instrumental view shows how geography can help us make progress toward some aspect of the real, but it does not tell us if we are making progress toward the good (or the bad), and I do not think anyone would say that moving toward the bad or evil is really progress.

Still, strong arguments can be made that no matter how much we wish otherwise, we cannot move beyond the instrumental—that all judgments are bound to particular projects and goals and while we may be able to expand the goals or projects, judgments will remain instrumental. I think this position is wrong. I believe that there are means of assessing progress that move us beyond particular projects. I refer here to the idea of intrinsic geographical progress. Let us consider what that is and how it is based on geographical conceptions of the good that are not relative to particular projects.

Intrinsic Geographical Progress

All places are real, but not all places are good. We should be careful here to understand that when we evaluate a place we are really evaluating aspects of it. But some aspects may be so important that they overwhelm the others, and so we can praise or condemn the place itself. The same of course is true of people. Even the worst person may have an attractive quality or two, which unfortunately are not sufficient to prevent the person from being condemned.

Nazi concentration camps, slave plantations, and violent homes are bad, even though a few instances of kindness may have taken place. Moral theories based on various conceptions of the good have been used to condemn these places. The Kantian categorical imperative and the utilitarian concept of the greatest good for the greatest number are cases in point. Instead of drawing on these theories to evaluate place, I will show how we can judge places according to qualities of the good that are illuminated by geography. I must stress here that the phrase *intrinsic geographical progress* (and also *intrinsic geographical judgments*) does not mean that these qualities are only good for geography. I use the phrase because I believe that these are qualities of the good and therefore are independent of geography but that geography holds them in a particular relation that illuminates them most clearly. These qualities of the good that geography draws upon allow us to judge whether places move us in the direction of both the real and the good and thus make real geographical progress.

Intrinsic geographical judgments and progress draw upon two qualities of the good. The first, discussed in the introduction, is the value of being as aware of reality as possible. It is good to be more than less aware and to see reality clearly. This good also asserts the converse: that what we call evil is largely the result of a lack of awareness—of not seeing clearly. The point is that few would intentionally inflict pain and suffering on others. If given the choice, people would prefer doing good.

The value of being aware and seeing reality clearly becomes expressed in geography as seeing as completely and publicly as possible how the world and its parts or places are interrelated. As geographical beings we are curious about the world. We want to know what is over the next hill, or what lies beyond the horizon. This curiosity about the world and its parts is expressed in the history of cartography, in every culture's efforts to map these relationships, even though these maps are often largely products of fantasy and mythology. Seeing the world as completely and realistically as possible is a public and democratic effort. It is possible only if we can share knowledge and compare views. It requires free and open access to information. Indeed, we can go farther and say that seeing clearly is an unselfish act, for it must be shared to know if it is true. And seeing clearly requires a complex social apparatus that promotes a free and open exchange of knowledge and provides everyone with opportunities to expand their horizons. We can call this first intrinsic geographical value *seeing through to the real*.

The second aspect of the good that geography draws upon is that we value a varied and complex reality. Part of the attractiveness of the real and the good is that they are inexhaustible and unfathomable. These qualities hold our attention. We want to become more aware of such a reality. But variety and complexity must to some degree be accessible to human understanding or else the world would be entirely alienating and our only recourse would be to escape from it. We believe that we know more about the world—that reality is clarified—when some of its complexity can be seen to follow from simpler sets of relations. We take these elementary relations that generate the others to be more basic, more essential, or more real. This clarification of the diverse and complex is a valued and unending part of intellectual life only if the complexity and variety are unending. This second aspect of the good as an intrinsic geographical value can be called *valuing a more varied and complex reality*, or *variety and complexity* for short.[3]

Seeing through to the real and valuing a more varied and complex reality are the two aspects of intrinsic geographical judgment that guide our geographical actions, and they must be applied jointly. It is good to create places that support variety and complexity in nature and culture, and it is good that varied and complex places in turn encourage us to want (and allow us) to expand our horizons and to know more. The two parts of intrinsic geographical judgments can be reinforcing so that they lead to an ever more complex and diverse set of places that increase our awareness of this expanding reality, and each judgment can set a limit on the other.

Seeing through to the real sets a limit on variety, for variety itself is not of value if it prevents us from understanding the real. Crack houses, opium dens, areas of poverty and deprivation, and places of superstitious custom can increase diversity or variety but are to be discouraged because they so severely diminish awareness of the real. And variety checks seeing through to the real, for if places are too transparent, if they are too much alike, or if nothing complex takes place in them, they do not offer enough difference and mystery to hold our attention.

The joint application of seeing through to the real and variety and complexity goes to the heart of place-making. Each and every place creates boundaries that offer some degree of opacity or transparency in order that activities may take place. We find opacity even in those places dedicated to heightening our awareness. A scientific research laboratory, for example, allows only those engaged in the research to enter. For everyone else, the place is not open to view. This opacity is justified in the long run if the results of the laboratory not only are made public but also expand our ability to see reality. The reasonable expectation that this will occur justifies the use of this place. In this way, the two facets of intrinsic geographical judgments jointly provide a guide that helps strike the right balance. Armed with intrinsic geographical judgments, we can now say why a place is or is not on the right track—and whether it can make intrinsic progress.

But it usually is not this simple. Consider, for example, the Iron Curtain of the Soviet Union. Stalin used its political boundaries to prevent people on the inside from seeing out and to prevent those on the outside from seeing in. He argued that this was necessary for Soviet security and the development of socialism. If we accept this as true, then it can be argued that instrumentally the Iron Curtain was a good or effective use of geography. It helped Stalin achieve these goals.

As noted above, moral relativists claim that instrumental judgments are the only kind. If we tried to justify instrumental judgment by some higher or more general values, they would say that these too were instrumental in that they were values of larger, more general projects. Stalin's project of socialism can be embedded in a broader goal of world socialism, or Stalin's concern about Soviet security can be embedded in a larger concern about national sovereignty. No matter how much more general our judgments became, relativists would argue that they still were judgments that pertained to projects. We cannot escape the instrumental loop; we can only extend it.

But applying jointly the two parts of intrinsic geographical judgments, we find that even though the Iron Curtain may have been an effective instrument for Stalin's purposes, it violated the value of seeing through to the real. It was part of reality, yet, and more importantly here, it strongly diminished our awareness of reality. How do the two balance out? I would argue that because of this extreme diminution of awareness, it prevented those inside and out from knowing the consequences of their actions and the possibilities of doing otherwise. It restricted awareness and choices, and for these reasons we can say that the Iron Curtain was not a move in the right direction.

Seeing through to the real and valuing a more varied and complex reality jointly provide a positive moral picture. The two suggest whether we are making moral progress. If we stress their opposites, we can also sketch a negative geographical landscape of where not to go. This landscape is dominated by three features: autarky, tyranny, and chaos.

Autarkic and secret places, of which the Iron Curtain is a large-scale example, are the result of inverting the value of seeing through to the real. Tyranny of one place over others to create a landscape of sameness violates the value of variety and complexity. And geographical chaos resulting from constant transgression of the rules of in and out of place prevents any project from being undertaken and violates both of the intrinsic values. These three types of geographical evil—autarky, tyranny, and chaos—provide insights into what is geographically objectionable about a Soviet Union or a Nazi concentration camp or an American slave plantation. Even though these may have made instrumental progress in their uses of geography, they moved us away from the direction of intrinsic geographical progress.

People often think that the major component of morality is justice. Justice is an important part of the moral, but theories of justice are not at the foundation of morality. Principles of justice depend ultimately on what is thought to be real and good. This is why our discussion begins at this basic level, but it too leads to a particular conception of justice and its stipulation in rights. Briefly, intrinsic geographical judgments recognize the value of reason and therefore the dignity of humans as reasoning creatures. The judgments require that there be free and open access to knowledge. This means freedom of the press, no government secrecy, and a commitment to democracy. Intrinsic geographical judgments stipulate that all persons be provided the means for expanding their intellectual horizons. This means that everyone should be allowed to develop to his or her fullest potential and that our responsibilities for others do not diminish with distance or political boundary. The judgments also recognize that poverty, disease, and malnutrition diminish our humanity; they narrow our horizons and thus should be eliminated. In other words, intrinsic geographical judgments lead to principles of justice and human rights that are recognized by the Bill of Rights in the U.S. Constitution and by the U.N. charter on human rights.

To sum up, then, real progress can be made as we move to the good. That means having our projects guided by intrinsic geographical values. Another way of putting this is to say that we make more progress when our instrumental judgments become more like our intrinsic ones. The guiding principles are the intrinsic geographical judgments. This suggests that we must be aware of them to make intrinsic progress. But let us consider how we can have an intuition about intrinsic progress even if we are not conscious that intrinsic judgments exist. We can do so because qualities of intrinsic judgments are latent in some aspects of instrumental progress. This latency allows us to build on our current practices and elevate them. We do not have to destroy all previous practices or erase past landscapes and places to make progress.

Excellence and Aesthetics as a Bridge to the Good

Instrumental progress provides the most accessible and universal experience of making progress in some sense. We find it available at any level and in any culture. All that is required is participation in a project. The experience does not make us think that the future will necessarily be better; nonetheless, by participating in projects and making progress instrumentally we may be able to sense some of the qualities that lead to intrinsic progress.

A child learns to read. If this is not the child's explicit goal, it is certainly a goal society assigns to him or her, and the child can make progress toward it. The child can progress slowly or rapidly, and his or her progress can be compared with that of other children who are attempting to master this task. In this case instrumental progress is acquiring a skill—a child learning a language or a game of chess, a graduate student mastering geographical theory, a baseball player becoming a better hitter, a violinist becoming a virtuoso, a man becoming a better parent.

The skills we acquire include using geography and constructing places for undertaking our projects. Indeed, instrumental judgments are assessments of such skills. The schools, homes, musical conservatories, and athletic training and playing fields can be built in ways that make the goals of the projects more or less easy to attain. The quality of the place can in fact extend the level of the project. With better schools we can learn more than we thought possible. The acoustics of our music halls can improve our music and push us to compose even better pieces. The same is true with athletics: new fields and training facilities have allowed athletes to reach new heights.

This is a sense of progress that we all have experienced and that we see in others. Doing a task well, seeing it done well by others, leads to several important virtues. We realize that in order to attain a goal we must discipline ourselves. We also must submit to a process. We may even have to learn to cooperate with others and help them in order to achieve the goals of the project. We learn to take pride in our work and yet recognize that others too deserve credit. And we can recognize that others may have more skill than we do, which can make us humble and yet spur us on. We can lose ourselves in our work and so provide ourselves a taste of what it is like to be part of something greater than ourselves, something that could even outlast us. If the project is done well or excellently and is of use to others, we feel we have provided a gift that they too can use and enjoy.

An important effect of these virtues is to move us closer to an aesthetic appreciation. Accomplishing things excellently means that we execute projects with a certain grace and economy of effort. These are aesthetic qualities of skill, and projects that are done skillfully have an aesthetic appeal. They are done gracefully and beautifully. Acquiring skills to undertake projects well can then increase awareness of the aesthetic qualities of our own efforts and those of others.

Aesthetic sensitivity is important as a bridge to the real and the good. In the *Republic,* Plato was concerned that beautiful things could be subverted—

that stories, poems, and myths could be used to lie and lead us away from the good—but he was clear that beauty itself is real and that that is good. Indeed, Plato thought that all things that are efficient are beautiful, and in the *Phaedrus* he tells us that beauty is the only form of the good that is accessible to the senses.

Kant too saw beauty as a bridge to the good or the moral. The empirical world, described in his *Critique of Pure Reason* as a world of unbroken causal networks, is closed in on itself—a heteronomous world, as he put it—and we as physical beings are part of that world. But we are also part of a moral world, described in his *Critique of Practical Reason* as a world of autonomy. Here we are free agents who must choose and who are responsible for our choices. Connecting these two worlds—the empirical world of heteronomy and the moral world of autonomy—is the power of judgment, which he described in his *Critique of Judgment*. Judgment takes the idea of purpose and plan that is part of the moral world and brings it to the empirical world by having the particular empirical case become a part of a more universal and intelligible system of relations. Judgment allows us to think of the particular as part of the universal. For Kant, aesthetic judgment was the most accessible means of illustrating this connection. An aesthetic judgment that something is beautiful not only focuses on the particular object and its qualities but also claims that the object is beautiful and is thus partaking of a universal quality. If the judgment were subjective, Kant would say the object was pleasing, but to call it beautiful is to make a claim that the beauty is real. And here the critical point is that although the object that is beautiful may seem to fit together in that its parts conform to the right proportions and symmetry, it does not itself fit or conform to any particular purpose. It is an end and not a means to an end. In Kant's view, beauty that has no further function is pure or intrinsic beauty, and an aesthetic judgment is pure only if the beauty of the object is an end in itself. Beauty then is like the good, for both are real and are ends in themselves. Yet beauty is more accessible than the good and can become an introduction to it. Like the good, it pleases immediately, apart from any purpose, project, or interest; and like the good, the beautiful rests on claims of universality. Thus, aesthetic judgments provide a means of sensitizing us to the more difficult issues of moral judgment. And even though pure aesthetic judgments are not governed by particular ends or goals of projects, our aesthetic sensibilities are cultivated by the skills we acquire as we engage in projects. That is, the aesthetic qualities of instrumental progress can make us more receptive to the qualities associated with the good and the moral.

Yi-Fu Tuan writes about the connection between the aesthetic and the moral. He defines the aesthetics as the senses come to life, as apposed to the anesthetic, which deadens our awareness and puts us to sleep. The aesthetic, like the moral, is expansive. It draws attention to the vastness and complexity of the world. It makes us aware of the general while still attending to the particular. It is a life-enhancing force. In *Passing Strange and Wonderful* Tuan says that "at a general level, good is simply a fusion of moral and aesthetic conceptions, which include a sense of rightness and appropriateness, care and accomplishment, delight in the way things are done and in the things (both natural and artificial) themselves."[4] Such sensibilities are cultivated by our acquisition of skills and in projects well done.

When instrumental progress produces these sensibilities, it can help provide a glimpse of the qualities of the real and the good in general and prepare us to move in that direction and make real progress. But we cannot rely only on instrumental progress. We must also think of the real and the good directly in the form of intrinsic geographical judgments. This is because the sensibilities that are cultivated by doing a task well may not be strong enough to move us beyond the realm of the task. We may appreciate the role of beauty in a project, but it may remain rooted in the context of that project alone. In Kant's terms, the aesthetic judgment remains impure, or in our terms, instrumental.

If we do not keep the real and the good in mind, excellence and the aesthetic may become subverted. They may be taken for the good itself and may even move us in the wrong direction. Someone engaged in a project that is evil, such as the commandant of a concentration camp, can still strive to be the perfect bureaucrat and find aesthetic qualities in many parts of his task. Nazi Germany was consumed by an aesthetic; we find evidence of this in the idea of the "perfect" Aryan, in state architecture, and in the "pure" German landscape. The aesthetic can often disguise evil. The beautiful and the sublime can be erected on the backs of injustices to other humans and to nature. So many grand aesthetic projects, such as Versailles, St. Petersburg, and the construction of innumerable palaces, gardens, and temples elsewhere around the world, were accomplished by displacing former inhabitants, by pushing slave and corvée labor beyond endurance, and by a callous disregard for the natural habitats. The radiance of these places is dimmed by the darker side of their histories.

Within the context of particular projects, then, excellence and the aesthetic by themselves are not enough to assure that we will move in the right direction. Excellence and beauty are neither substitutes for the good nor sufficient to lead us to it. Even if they may make us more receptive to the good, we still need a conception of it to move in the right direction. But as ideals that transcend any particular project, excellence and the aesthetic share the same qualities that we find in intrinsic geographical judgments: a dedication to opening our eyes to the real and valuing the complexities of the world.

The aesthetic impulse is not the only means of moving from the instrumental to the intrinsic. A curiosity about the world that is spurred by science can move us in the direction of the real and the good. Indeed, many see science as the road to the real, for disclosing the real is science's avowed intention. Of course science can be employed as part of a project, but the science I am talking about here is the discipline of science in general, which is far broader than any project. Like anything else that is done well, science is aesthetic. Its means of revealing the real depends mostly on its methods. These make science a self-critical enterprise, constantly comparing its models with the real. Scientists are convinced that they have made progress in identifying the real, and those branches of geography that deal primarily with the natural world may have made progress in this intrinsic sense. They can move us closer to the real, and this of course is good.

But when we consider not only the physical world but also the human world—that part of reality that is created by human activity—we find that progress is more difficult. Part of the reason lies in the dual facets of geography—our role as agents and our intellectual awareness of the world. This duality means that we need not only to make progress in understanding the world in an intellectual sense but also to create a world that allows us to do so and thus make progress materially through the places we construct. The combination of being agents, being aware of ourselves as agents, and having our agency affect our awareness makes progress all the more difficult.

Yet the intrinsic geographical judgments still point to the direction we need to go and enable us also to recognize places that have already taken us partway there. Certainly schools and universities when adhering to their mission are places that attempt to increase our awareness and also value more complexity than less. Even the act of teaching itself is a gift that when well given is nothing less than insight into the real and the good. We must encourage these institutions and practices and make sure they do not become overwhelmed by other issues or become instruments of other projects. We also can assume that all endeavors to know the world truly require free and open access to information, and so places that encourage this and that discourage censorship and secrecy are essential for the real and the good. And we know that poverty and illness sap our strength and prevent us from expanding our horizons, so we should support places that value human rights and dignity. The point is that the intrinsic criteria for judging provide a means of assessing these places and a direction to the real and the good and allow us to make real progress.

How far can we go? How close can we get to a heaven on earth? There are of course innumerable impediments that, as in all discussions of morality, make failure and evil always more vivid and more easily imagined than success or the good. But two difficulties should be mentioned. One is that the criteria of intrinsic geographical judgments create their own tensions, which cannot be resolved beforehand. They can provide checks on each other, but these are not amenable to formula. We have little way of knowing how well we can balance seeing through to the real and variety and complexity unless we try. The moral judgments, then, offer only guideposts to show us the way. They do not provide pronouncements or commandments that encapsulate the good. This presents the second and more general difficulty in making progress: although real, the good is ineffable. It cannot be known in its entirety. Although facets of it can be discernible, as a whole it cannot be anything else but a suggestion, a hint. It acts as a beacon or magnetic attraction, beckoning us on. Is this enough? It must be. If it were to be completely knowable and tangible, then it would become far too limiting and rigid to be thought of as real and objective and deserving of our highest consideration.

For Yi-Fu Tuan this ineffable quality provides, as he puts it, "a powerful lure which limits the indecisiveness of freedom and yet does not enslave or blind." Recognizing this ineffable quality is extremely important today for it presents us, again in his words, with "a safeguard against intolerance and moral stasis."[5]

Notes

This work draws upon and extends my article "A Sketch of a Geographic Theory of Morality," *Annals of the Association of American Geographers* 55 (March 1999): 26–44. It also repeats in part the discussion in Robert Sack, "The Geographic Problematic: Moral Issues," *Norwegian Journal of Geography* 55 (September 2001): 117–25.

1. I am a realist. (Actually I should say that I am a *critical* realist, rather than a *naïve* one, because no one now can be anything without also claiming to be aware of the foundations and limitations of a position.) This means that I, along with virtually all other realists, assume that there is a reality and that we must attempt to understand, symbolize, model, or represent it. It also means that I accept that we can never know all of it or even know it as it "really" is, or that we can describe clearly how we gain confidence that some models more closely correspond to reality than do others, but I do not accept that any model is as good as any other. Nor do I accept that reality, though existing, has no effect on us. Not only are we influenced by it, we are drawn intellectually to it. I argue here that the same applies to the good.

2. Saying that place has an effect and helps us do things is not to fetishize place. Rather it is a realistic recognition that place enables us, and does so by acting in part as an agent. But places cannot exist and operate without us. These mutually constitutive relationships are part of the relational system of place and self that is explored in Robert Sack, *Homo Geographicus: A Framework for Action, Awareness, and Moral Concern* (Baltimore: Johns Hopkins UP, 1997).

3. I do not pretend to know how to measure variety and complexity. Even biology does not have a definite measure, and yet biologists make claims that biodiversity is good.

4. Yi-Fu Tuan, *Passing Strange and Wonderful* (Washington, D.C.: Island Press, 1993), 214.

5. Yi-Fu Tuan, *Morality and Imagination: Paradoxes of Progress* (Madison: U of Wisconsin P, 1989), 180.

Robert David Sack

Contributors

J. Nicholas Entrikin is professor of geography at the University of California, Los Angeles, and the author of numerous publications, including *The Betweenness of Place: Towards a Geography of Modernity* (Johns Hopkins UP, 1991) and, more recently, works on place in democratic theory, among them "Political Community, Identity, and Cosmopolitan Place," *International Sociology* 14 (1999), and "Le langage géographique dans la théorie démocratique," in *Logique de l'espace, l'esprit des lieux: Géographies à Cerisy*, ed. J. Lévy and M. Lussault (Paris: Belin, 2000).

David Lowenthal is professor emeritus of geography at University College London and the author of numerous books, among them *West Indian Societies* (Oxford UP, 1972), *The Past Is a Foreign Country* (Cambridge UP, 1986), *The Heritage Crusade and the Spoils of History* (Viking, 1996), and *George Perkins Marsh: Prophet of Conservation* (U of Washington P, 2000), which received the 2001 J. B. Jackson Prize of the Association of American Geographers.

Kenneth R. Olwig is professor of geography at the University in Trondheim and the author of *Nature's Ideological Landscape* (Allen & Unwin, 1984) and *Landscape, Nature, and the Body Politic* (U of Wisconsin P, 2002). He is also co-editor of *Nordscapes: Landscapes and Regional Identity on Europe's Northern Edge* (U of Minnesota P, 2002).

Robert David Sack is Clarence J. Glacken and John Bascom Professor of Geography and Integrated Liberal Studies at the University of Wisconsin–Madison and the author of *Conceptions of Space in Social Thought* (U of Minnesota P, 1980), *Human Territoriality: Its Theory and History* (Cambridge UP, 1986), *Place, Modernity, and the Consumer's World: A Relational Framework for Geographical Analysis* (Johns Hopkins UP, 1992), and *Homo Geographicus: A Framework for Action, Awareness, and Moral Concern* (Johns Hopkins UP, 1997).

Yi-Fu Tuan is the John K. Wright and Vilas Professor of Geography, Emeritus, at the University of Wisconsin–Madison and the author of fifteen books, most recently *Morality and Imagination: Paradoxes of Progress* (U of Wisconsin P, 1989), *Passing Strange and Wonderful: Aesthetics, Nature, and Culture* (Island Press, 1993), *Escapism* (Johns Hopkins UP, 1998), and *Who Am I?* (U of Wisconsin P, 1999).

Thomas R. Vale is professor of geography at the University of Wisconsin–Madison and the author, co-editor, and editor of eight books, including *Plants and People: Vegetation Change in North America* (Association of American Geographers, 1982), *Progress Against Growth: Daniel B. Luten on the American Landscape* (Guilford, 1986), *U.S. 40 Today: Thirty Years of Landscape Change in America* (U of Wisconsin P, 1983), and *Walking with Muir Across Yosemite* (U of Wisconsin P, 1998).

Index

absolutism, 29, 37, 59n65, 115
Adams, Brooks, 65
adaptationism, 2–4, 6, 10–12, 14, 15, 18
adaptation/progress model, 16–18;
and atmospheric processes, 14–15; of
geomorphology, 11, 13; of soil development, 6, 7, 9; of vegetation dynamics, 3
adventure, 85–86
aesthetics: and the good, 124–29; and
progress, 89–90, 116
agriculture, ix, 40, 81–82, 91
Albion's Triumph (masque), 33, 38–39
Allen, T. F. H., 4
America: anxiety in, 78, 91–93; artifice
in, 83; Bill of Rights of, 124; community in, 98–99; environmental-
justice movement in, 52; history in,
65, 91, 94; liberal education in, 103–
7; political landscape of, 26
American Revolution, 73
Anhalt-Dessau (Germany), 40
animals, 6, 88, 96n19
Anne of Denmark (queen of James I of
England), 29, 30, 31
Antigone (Sophocles), 84
anxiety: in America, 78, 91–93; and architecture, 82–83, 93; and awareness,
93–95; and culture, 68, 79–80; ecological, 61, 62, 68; about the future,
65, 66, 67, 80; in media hype, 69–

70; and morality, 88–89; and nature,
80, 83, 84–85, 87–88, 91; and progress, xiii–xiv, 66, 68–70, 78–96
architecture: and destruction, 45,
59n57; and fear of progress, 82–83,
93; and landscape parks, 59n62;
modern, 45–50; Nazi, 127; Palladian,
40, 46, 49, 57n39; and perfection,
98; and utopianism, 62
Arendt, Hannah, 98
aristocracy, 102, 104–5
Aristotle, 54n4
atmospheric processes, xi–xii, 2, 11, 12,
14–18
automobiles, 48–49, 58n55, 59n60, 91
awareness: and aesthetics, 125–26, 127,
128; and agency, 128; and anxiety, 93–
95; of complexity, 129; and education,
128–29; and globalization, 94; and
morality, 94–95, 102, 109, 114–15; of
reality, xv, 114–15, 117–18, 120–23

Bacon, Francis, 36, 45, 50
Banham, Reyner, 62
beating the bounds, 27, 28
beauty. *See* aesthetics
Bellamy, Edward, 62
Berman, Marshall, 45
Bloom, Allan, 106
Botkin, D. B., 1, 19n10, 20n12
bourgeoisie, 43–44

Boyle, Richard, 40
Brahe, Tycho, 29
Brand, Stewart, 73
Britain, 26–27, 37; concept of, 30, 33, 34; unification of, 29, 38, 39. *See also* England
Brunner, Otto, 29
Buddhism, 87
Bunyan, John, 41

Calvin, John, 82
Canterbury Tales (Chaucer), 41
Carlyle, Thomas, 65
Cavell, Stanley, 101–3, 105, 109
change, xiv, 17, 18, 79, 113; cultural, 71–72; and custom, 27–28; cyclical, xii, 80; and education, 108–9; in elements of place, 118–19; environmental, 69, 71–72, 73; and the future, 63, 67–68, 69, 71; historical, viii–ix; linear, xiii, 3; and religion, ix, 109
Chaucer, Geoffrey, 41
China, 80–82, 84, 87
Christensen, N. L., 4
Christianity, ix–x, 45, 64, 87, 100–101
Christian IV (king of Denmark), 29
chronosequence model, 6
The City of Tomorrow (*Urbanisme;* Le Corbusier), 46–51, 53
civil rights movement, 92
class, ix, 85; aristocratic, 102, 104–5; bourgeois, 43–44; and education, 91–93; landed gentry, 39
class struggle, 44
Clements, Frederick, 3, 4
climate. *See* atmospheric processes
Closing of the American Mind (Bloom), 106
Coke, Edward, 27, 36
Communism, 43–45
Communist Manifesto (Marx), 43–44
community: and architecture, 98; and

country, 56n19; Dewey on, 105–6; and law, 25, 27, 52; and marching, 54n7; and nature, 108–9; organic, 54n7; and outsiders, 73, 87; and perfection, 108, 109, 110; and place, 22–23, 52, 54n6, 97–98; and religion, 54n6; and republicanism, 104, 107; and self, 97–98, 100, 108; and state, 24; and stewardship, 72–73; words for, 55n11. *See also* social systems
computers, 84, 91, 95
Condorcet, Antoine-Nicolas De, 42–43
Confucianism, 81
contingency, 5–6, 84–85. *See also* history
Conversations on the Plurality of Worlds (Fontenelle), 42
Copernicus, 89
counter-culture movement, 62
country, 55n8, 56n19; court *vs.*, 26–30
Crèvecoeur, St. John de, 99
Critique of Judgment (Kant), 126
Critique of Pure Reason (Kant), 126
culture: and anxiety, 68, 79–80; change in, vii–viii, 71–72; consumer, 98, 100; minority, 92–93; and nature, 68, 108, 116, 118, 121; and perfectionism, 101, 109; and reality, 117
custom, 37, 79, 86, 101, 122; glorification of, 92–93; and law, 27–28, 36, 52; and political interests, 50, 52; and progress, 43, 118

Dagger, Richard, 103–4
Daniel, Samuel, 30
Darwin, Charles, 44
Davis, William Morris, 10, 11, 13, 17
Dee, John, 57n29
deities, 82–83, 85, 86, 93, 101. *See also* religion
democracy, 50, 52, 109, 121, 124; *vs.* aristocracy, 104–5; Cavell on, 102–3; and

the future, 69, 72; and liberal education, 103–7; and morality, 103, 107,
108; and perfectibility, xiv–xv, 97–
112; perfectionism in, 98–99, 100,
102; and place, 99, 107–8; republican and liberal models of, 103–4, 107
Democracy in America (Tocqueville), 99
determinism, 3–4, 7, 10, 12, 14, 15;
Gould on, 2, 18
Dewey, John, 105–6, 109
Durkheim, Emile, 54n7, 72

economics, 66, 116
Edison, Thomas, 91
education, xiii, 70, 95; and awareness,
128–29; and class, 90, 91–93; classical, 105; and democracy, 103–7; liberal, 97, 99–100, 103–7; and minorities, 94; moral, 100, 102, 109; and
perfectionism, xv, 100, 108, 109; and
place, 116, 117, 118–19; postmodernist, 104–5
egalitarianism, 91, 103
Egler, F. E., 4
Elizabeth I (queen of England), 24, 25,
26, 37
Emerson, Ralph Waldo, 101, 102–3, 105,
108–9
enclosure, 40
Engels, Friedrich, 44, 45, 46
England, 24–31, 34, 39, 40, 54n8, 78.
See also Britain
the Enlightenment, 18, 61, 62–63
Entrikin, J. Nicholas, xiv–xv, 97–112
environment: anxiety about, 61, 62, 68;
change in, 69, 71–72, 73; and community, 108; damage to, 65–66, 67,
68; Dewey on, 106; and political
landscape, 27; and soils, 6–7, 9; and
vegetation dynamics, 3, 4, 6
environmental-justice movement, 52
Europe, 65, 81–82. *See also* the West

evolution, xi, 2, 18, 44
existentialism, 101

family, xiv, 86–87, 88, 116; dysfunc-
tional, 119, 120
Fawkes, Guy, 54n8
Five Weeks in a Balloon (Verne), 62
Flaubert, Gustave, 64
Florimène (play), 32
Florio, John, 55n11
Fludd, Robert, 57n29
Fontenelle, Bernard le Bovier de, 42
The Fortunate Isles, and their Union
(masque; Jonson), 38
Franz, Prince (Anhalt-Dessau, Germany), 40
French Revolution, 64, 65
the frontier, 91–92
Frye, Northrop, 37–38
fundamentalism, 107
the future, 61–77; anxiety about, 65, 66,
67, 80; and change, 63, 67–69, 71;
immediate, 72; and instrumental
progress, 124; and the past, 64–65,
72; positive attitudes towards, 71–74;
and the present, 64; private, 67–68

geography: definition of, vii–viii; and
the good, 120, 121; human beings as
agents of, vii, ix, x, xi, 117, 121, 128;
physical, xi, 1–21; and progress, vii,
113–30; and reality, 128
geomorphology, xi, 2, 9–13, 16–18
Ghil, M., 15
Glorious Revolution (England; 1688),
39
the good, 113–30; and aesthetics, 124–
29; awareness of, 114–15; ineffability
of, 115, 129; and progress, 124, 127;
and reality, 113, 119, 120, 129. *See also*
morality
Goodlett, J. C., 3

The Good Life (Tuan), 97
Gould, Stephen Jay, xi, 1–3, 5, 10, 18
gradualism, 3–4, 7, 8, 10–12, 14, 15;
    Gould on, 2, 18
Graf, W. L., 12
Greece, 80, 84, 100–101, 104

Hack, J. T., 3
Hegel, G. W. F., 64
Heidegger, Martin, 101, 102
Henry, J. D., 4, 5
Hiroshima, 65
historical-event model, 16–18; and at-
    mospheric processes, 14–15; of geo-
    morphology, 11–13; of soil develop-
    ment, 7–9; of vegetation dynamics,
    5–6, 9
history: accelerated pace of, 64–65; in
    America, 65, 91, 94; and change,
    viii–ix; destruction of, 50, 59n55, 124;
    Dewey on, 106; end of, 44; and guilt,
    70; preservation of, 52, 65, 66, 71–
    72; and religion, 63, 64; and totalitar-
    ianism, 98
Hobbes, Thomas, 34, 35, 36
Hobsbawm, Eric, 27–28
Holocaust, 114, 119, 120, 123, 127
Holy Roman Empire, 64
Hutchins, Robert M., 106

imagination, viii, 79
immediacy, tyranny of, 69–70
imperialism, 65
the individual, 67–68, 73–74, 98, 103,
    105, 110
industrial revolution, 65
industry, 40, 70
instrumental geographical judgments.
    See under judgments
International Exhibition of Decorative
    Art (Paris; 1925), 48, 58n55
the Internet, 95

intrinsic geographical judgments. *See
    under* judgments

James I (king of England), 24, 26–31,
    34, 40, 54n8
Jenny, Hans, 6, 9
Johnson, D. J., 7
Jones, Inigo, 30–34, 36–38, 40, 41, 46,
    49, 57n29
Jonson, Ben, 30, 36, 38, 41
Josselin, Ralph, 84–85
Judaism, ix, 45, 82, 100–101
judgments: aesthetic, 126; instrumental
    geographical, xv–xvi, 115–16, 124,
    125, 128; intrinsic geographical, xv–
    xvi, 115–16, 120–24, 127–29; moral,
    126, 129. *See also* progress: instru-
    mental geographical; progress: in-
    trinsic geographical

Kant, Immanuel, 120, 126, 127
Kauffmann, Stanley, 83–84
Kent, William, 40, 57n39
knowledge, 95, 97, 99, 121, 124; limits
    of, 93, 115
Knox, J. C., 12, 13

landforms. *See* geomorphology
landscape: destruction of, 124; *vs.* lord,
    26–30; meanings of, xii, 29, 56n19;
    Nazi, 127; and perfectibility, xiv–xv,
    98, 99; and place, 22–60; political,
    24–27; social, 24; and the state, 30–
    36, 41–42; in theater, 25, 36, 38, 40;
    and utopianism, 52
landscape garden parks, 39–40, 57n39;
    Le Corbusier on, 49, 59n62
Landschaft (landscape), 29, 56n19
Last Judgment, 64
law: common, 27–28; and community,
    25, 27, 52; customary, xii, 27–28, 36,
    52; and heritage protection, 65; and

landscape, 25–26; natural, 29; and perfectionism, 101; and progress, 24, 90; statutory, 29, 36
Le Corbusier, 45–50, 51, 53, 59nn57,62,65; on automobiles, 48–49, 58n55, 59n60
L'Enfant, Pierre, 99
Letter to D'Alembert (Rousseau), 108
Leviathan (Hobbes), 34, 35
liberalism, 103–4, 107
Looking Backward (Bellamy), 62
Louis XIV (king of France), 50, 51, 59n65
Lowenthal, David, xiii, 61–77

Machiavelli, Niccolò, 37, 63
Malthus, Thomas, 67
Marx, Karl, 43–45, 46, 101
The Masque of Beauty (Jonson), 38
The Masque of Blackness (Jonson), 30–31
Mayer, Martin, 78
McCune, B., 4
McFadden, L. D., 9
media, xiii, 69–70, 73
Mencius, 81
metropolitan frontier, 91
A Midsummer Night's Dream (Shakespeare), 37
minorities, 91–93, 94
modernity, ix, x–xi, xii, xiii, 24; in architecture, 45–50; dialectic of, 23, 28, 43–45, 52; and education, 104–5; and nature, 87–88; and science, x, 46
monarchy, 24, 25–26, 81; absolute, 29, 37, 59n65, 115
Montesquieu, Charles de Secondat, baron de La Brède et de, 40
morality: and aesthetics, 127; and architecture, 98; and awareness, 94–95, 102, 109, 114–15; and custom, 86; and democracy, 103, 107, 108; and education, 100, 102, 109; impediments

to, 129; and judgment, 126, 129; and perfectionism, 101–4, 108, 109; in physical geography, 7, 10, 15–16; and place, 114; in political theory, 99; and progress, viii, xiv–xv, 68, 86–89, 97, 109, 115, 116, 118, 123; relativistic, 122–23; and rights, 123–24; theories of, 120. *See also* the good
Mormons, 73
Mo-tzu, 81, 87
Munz, Peter, 63
music, 52–53
Musset, Alfred de, 64
myth, 79, 80, 121, 126

Namias, J., 14
nature, xi–xii, xiii, 1; and aesthetics, 128; and anxiety, 80, 83, 84–85, 87–88, 91; and community, 108–9; complexity of, 130n3; and culture, 108, 116, 118, 121; and law, 29; and psychology, 97–98; and technology, 65–66; and theater, 42; transformation of, vii–viii, 66, 81; and utopianism, 62
The New Atlantis (Bacon), 36–37, 45, 50
Newton, Isaac, 90
New York, 49, 62
Nietzsche, Friedrich, 101
El Niño, 14–15

Oakeshott, Michael, 97
Oberlander, T., 11
Olwig, Kenneth R., xii, 22–60
Orgel, Stephen, 39
Owen, H. P., 100
Oxford English Dictionary, 41

Palladio, Andrea, 46, 49, 57n39
Pangle, Thomas, 104–5
papacy, 64
Paris, 45, 47, 48, 49
Park, Robert, 105

Parliament, English, 24, 27, 28, 39, 40, 54n8
Pascal, Blaise, 89 90
Passing Strange and Wonderful (Tuan), 99, 108, 127
perfectibility, xiv–xv, 64, 97–112
perfectionism: Cavell on, 102–3; and community, 108, 109, 110; and democracy, 98–99, 100, 102; and education, xv, 100, 108, 109; in geomorphology, 10; and morality, 101–4, 108, 109
perspective, artistic, 30–33, 36, 38, 42, 56n19
Phaedrus (Plato), 126
Phillips, J. D., 9
photography, aerial, 47, 48
physics, xii, 16, 17, 18
pilgrimage, 22, 41, 54n7
Pilgrim's Progess (Bunyan), 41
place, 22–60; and change, 118–19; and community, 22–23, 52, 54n6, 97–98; construction of, 54n6, 114, 122; and democracy, 99, 107–8; Dewey on, 105; and education, 116, 117, 118–19; and progress, 115, 116–20, 125, 128, 129; and projects, 117–20, 123; and the self, 107, 108, 117, 130n2; sense of, 22–23, 26, 52, 53n1, 54n4
Plato, 37, 45, 101, 102, 106, 125–26
Plumb, J. H., 78
politics, 50, 52, 99, 103, 116; and landscape, 24–27
populism, 69
postmodernism, 105
pragmatism, 108–9
the present, 64, 71–72; tyranny of, 69–70
progress: circuitous, 23–24, 25, 41, 43, 52, 53, 54n4; destruction for the sake of, 45, 50, 52, 59n57, 93–94; geographical, vii, 113–30; and hubris,

82–84; and institutions, 86; instrumental geographical, xv–xvi, 115–20, 122–23, 124, 125, 127, 128; intellectual, 89–91; intrinsic geographical, xv–xvi, 115–16, 118, 120–24; linear, xii, 2, 23, 25, 41, 43; march of, 23, 24, 43; material, 81–84, 86; meanings of, xii, 40, 41–42; modern views of, x–xi, xii; negative intrinsic, 123; personal, 84–86; and projects, viii, x–xi, xv, 117–20, 123, 124–28; topian, 23
Progress: Fact or Illusion? (Marx and Mazlish), 61
psychology, 97–98
Ptolemaic model, 89
Ptolemy, 37
punctuated equilibrium, 18
Putnam, Hilary, 102
Pythagoras, 37, 53

Rawls, John, 104
Reagan, Nancy, 85
realism, critical, 130n1
reality, vii–viii, 113–30; and aesthetics, 125–26; and autarky, 123; awareness of, xv, 114–15, 117–18, 120–23; complexity and diversity of, 121–22, 123, 128, 129, 130n3; cultural, 117; geographical, 117, 128; and the good, 113, 119, 120, 129; and instrumentalism, xv–xvi, 127
reciprocity, 86–87, 88
relativism, 105, 113, 115, 118, 122–23
religion, x, 37, 41, 54n6, 87–88; and change, ix, 109; fundamentalist, 70; and history, 63, 64; and perfectibility, 100–101. *See also* Christianity; Judaism; myth; ritual
the Renaissance, 24, 36, 42, 50
Republic (Plato), 45, 101, 102, 125–26
republicanism, 104, 107
revolution, ix, 43–45

Rhodes, B. L., 11
Richmond Park (England), 28
rights, human, 92, 123–24, 129
ritual, 37, 39, 79; circuits in, 23–24, 25, 27; sacrificial, 82–83
Rorty, Richard, 105, 106
Rousseau, Jean-Jacques, 64, 108, 109
Rubens, Peter Paul, 34
rurban-cybernetic frontier, 91
Rygg, Kristen, 57n33

Sack, Robert David, vii–xvi, 107, 113–30
St. Augustine, 101
St. Thomas Aquinas, 101
Sandel, Michael, 106–7
Scandinavia, 25–26
Schumm, S. A., 11, 12
science, xi, 1, 2, 42, 62, 128; and class, 91–93; and education, 104, 105; as frontier, 92–93; and history, 64; limits of, 66–67, 68, 69; and magic, 57n29; mistrust of, 70; and modernity, x, 46; and the state, 29, 37, 39
science fiction, 62, 67
Scotland, 26, 34, 39
the self: and community, 97–98, 100, 108; and democracy, 103, 107; duality of, 109; and place, 107, 108, 117, 130n2; sense of, 102–3
Shakespeare, William, 37, 41
Shang dynasty (China), 81, 82
Sketch for a Historical Picture of the Progress of the Human Mind (Condorcet), 42–43
slavery, 104, 114, 119, 120, 123, 128
socialism, scientific, 44
social justice, 41, 104, 107–8. *See also* rights, human
social systems, xii, 69, 85, 118–19. *See also* class; community
Socrates, 106
soil genesis, xi, 2, 6–9, 16–18

Sophocles, 84
Ssu-ma Ch'ien, 81
stages, 41, 42, 58n45
Stalin, Joseph, 122–23
Stapledon, Olaf, 62
the state, 52, 99; and landscape, 41–42; and progress, xii, 24–25, 89; sciential, 37, 39; territorial, 36; and theater, 25, 30–36, 37
state-factor equation, 6, 9
Steiner, David, 108
stewardship, xiii, 72–73
Stoicism, 87, 101
Swan, J. M. A., 4, 5

Taoism, 81
technology, ix, x, 83; and anxiety, 65, 66, 67, 70; and awareness, 94–95; and charity, 87–88; and class, 91–93; limits of, 68; and luck, 84–85; unintended consequences of, 67; and utopianism, 62, 63. *See also* science
theater, xii, 26–27, 42; gardens as, 57n39; landscape in, 25, 36, 38, 40; and the state, 25, 30–36, 37
Third World, 66
Thoreau, Henry David, 54n7, 101, 108
time, ix–x; scales of, 14–15, 17, 18; and soil development, 6–7
Time Vindicated to Himself and to His Honours (masque), 34
Titanic (ship), 83–84
Tocqueville, Alexis de, 99, 103, 104–5
topianism, 50–53
totalitarianism, 98
tradition, ix, 44, 50, 62; *vs.* custom, 27–28
transcendentalism, 108–9
Tuan Yi-Fu, vii, xiv, xvin1, 78–96; on aesthetics, 127; on delight, 73; on democracy, 99, 108; on education, 97, 109; on the good life, 97, 102; on in-

Tuan Yi-Fu (*cont.*)
effability of the good, 129; on marching, 23; on place, 22, 53; on progress,
18
Turkey, 87–88
Turner, Victor, 54n7
Twelfth Night (Shakespeare), 37
*Two Historical Accounts of the Making of
the New Forest and Richmond Park*, 28

United States. *See* America
universalism, xii, 81, 87–88
urban development, 62
Urbanisme (The City of Tomorrow; Le
Corbusier), 46–51, 53
urban-manufacturing frontier, 91
utilitarianism, 120. *See also* judgments:
instrumental geographical
utopianism, 30, 61; and march of progress, 23, 24; and revolution, 44, 45;
and Stuart state, 37–39; and technology, 62, 63; *vs.* topianism, 50–53

Vale, Thomas R., xi–xii, 1–21
vegetation dynamics, xi, 2, 3–6, 9, 16–

18; successional sequences in, 3, 4, 6,
10
Verne, Jules, 62
Vision of the Twelve Goddesses (masque;
Daniel), 30–31
Voisin Plan (Le Corbusier), 47, 48, 49,
58n55

Wakefield, Thomas, 28
Walpole, Horace, 57n39
Walzer, Michael, 98
Watson-Stegner, D., 7
weather. *See* atmospheric processes
Weinberg, Steven, 90
Weldon, R. J., 9
Wells, H. G., 62
the West, 80–83, 93
When the Sleeper Wakes (Wells), 62
Wilde, Oscar, 84
Womack, W. R., 11, 12
women, rights of, 92
World War II, 59n55

Yates, Frances, 57n29